Microbiology of African Fermented Foods

Fermented food products play a unique and critical role in the daily lives of inhabitants of the African continent, serving as a primary source of diet, providing nutrients and supplying health-promoting benefits, with some having cultural and spiritual connotations. Fermentation is used in a wide range of food and beverage applications, and the technology for enhancing this process is continually evolving. This book provides a broad yet succinct insight into the role of microorganisms in African food products and systems, with chapters dedicated to wider classifications of these fermented foods. It offers new perspectives and covers recent developments regarding the microbiological complexity of these food products. This book would be a valuable resource for students, food technologists, microbiologists, scholars and policymakers highlighting the contribution to the discourse on indigenous fermentation science and African food innovation.

Microbiology of African Fermented Foods

Gabriel Bidemi Akanni,
Abdullahi Adekilekun Jimoh
and Oluwafemi Ayodeji Adebo

CRC Press
Taylor & Francis Group
Boca Raton London New York

CRC Press is an imprint of the
Taylor & Francis Group, an **informa** business

Designed cover image: G.B. Akaani, A.A. Jimoh, and O.A. Adebo

First edition published 2026
by CRC Press
2385 NW Executive Center Drive, Suite 320, Boca Raton FL 33431
and by CRC Press

4 Park Square, Milton Park, Abingdon, Oxon, OX14 4RN

CRC Press is an imprint of Taylor & Francis Group, LLC

ISBN: 9781041111085 (hbk)
ISBN: 9781041111054 (pbk)
ISBN: 9781003658320 (ebk)

DOI: 10.1201/9781003658320

Typeset in Adobe Caslon Pro
by KnowledgeWorks Global Ltd.

Contents

Preface

Fermentation has been at the core of African culinary tradition for ages, not just as a means of food preservation but as a powerful expression of culture, identity and innovation. Across the continent, diverse fermentation practices have evolved, shaped by the richness of local environments, ingredients and indigenous knowledge. The fermented foods and drinks such as *ogi*, *kenkey*, *injera*, *dawadawa*, *pito*, palm wine, *khlii*, *tonto* and *marula* beer are not only a staple for millions of Africans but also a catalogue of complex microbial ecosystems that hold vast nutritional, functional and economic potential.

This book, ***Microbiology of African Fermented Foods***, looks at the multifaceted role of microorganisms in traditional African food systems. It is a scientific yet culturally grounded examination of how bacteria, yeast and other microbes turn indigenous African crops and raw materials into nutrient-rich, safe and tasty products. It highlights the different types of fermentation such as lactic acid, alkaline and alcoholic, and provides a regional breakdown of fermented foods to show how fermentation practices differ and intersect across West, East, Central, Southern and Northern Africa.

The purpose of this book is threefold: firstly, to document the vast diversity and microbiological complexity of African fermented foods; secondly, to provide insights that bridge traditional knowledge with modern science; and thirdly, to inspire further research, innovation and value addition in African food systems. Additionally, this book aims to preserve heritage while unlocking future opportunities for food security, safety and sustainability in Africa.

This book would be a valuable resource for students, food technologists, microbiologists, scholars and policymakers highlighting the contribution to the discourse on indigenous fermentation science and African food innovation.

About the Authors

Dr. Gabriel Bidemi Akanni is an accomplished food scientist and microbiologist specialising in food microbiology, biotechnology and food safety. He earned his PhD in Food Science from the University of Pretoria, South Africa. His research interests span antimicrobial resistance, microbial risk assessment and the use of molecular and genomic techniques for pathogen surveillance. Dr. Akanni has held postdoctoral fellowships at the University of Johannesburg and tenure-track position at Mountain Top University, where he contributed to food-borne disease epidemiology surveillance projects, co-funded by international organisations such as the Bill and Melinda Gates Foundation. He has published extensively on fermented food products, microbial safety in food production and the metabolomics of African fermented foods. His work has been presented at international conferences and has garnered him several prestigious awards, including a VLIR-UOS scholarship and sponsorship from the International Committee for Food Microbiology and Hygiene. Currently, Dr. Akanni continues to contribute to food science and biotechnology through research, teaching and collaborative projects across Africa and globally.

Dr. Abdullahi Adekilekun Jimoh is an accomplished researcher and microbiologist specialising in industrial and environmental biotechnology. He earned his PhD in Microbiology from the University of KwaZulu-Natal (Westville Campus), South Africa, in 2020, where his doctoral research focused on biosurfactant synthesis, product optimization, biotechnological applications, and molecular perspectives. Dr. Jimoh possesses extensive experience in a diverse range of laboratory and analytical techniques, including microbiological methods, molecular biology (gene cloning, expression studies, and gene knockouts), genetic engineering, bioinformatics, bioprocess development and various omics approaches. He has contributed to several research projects supported by prestigious organizations, including the National Research Foundation of South Africa, North-West University (NWU) Postdoctoral Grant, Council for Scientific and Industrial Research/Technology Innovation Agency/Department of Science Innovation (CSIR/TIA/DSI South Africa), the South African Water Research Commission (WRC), University of Johannesburg (UJ) and the Faculty of Science Research Committee. Dr Jimoh remains an active member of several professional scientific associations and ccurrently, he contributes to microbiology and biotechnology through research, teaching, supervision, and collaborative projects across Africa and globally.

Prof. Oluwafemi Ayodeji Adebo is Vice Dean: Global Engagement (Faculty of Science) and Professor of Food Technology and Director of the Centre for Innovative Food Research (CIFR) at the University of Johannesburg (UJ), South Africa. His research and activities at the Centre focus on improving traditional food processes, enhancing food-health-promoting food constituents, developing nutrient-dense food products and valorising of underutilised food sources using conventional, novel, and innovative approaches, including 3D food printing – all an extensive publication record of about 200 scholarly contributions including articles in high-impact journals, book chapters, reports and edited books. His research excellence has been recognized through prestigious awards, including the Institute of Food Technologists Emerging Leader Award, the National Research Foundation Emerging Researcher of the Year, UJ's Vice Chancellor's Distinguished Award as an Outstanding Young Researcher, and multiple UJ Faculty of Science Top Research Awards. He is a Young Affiliate of The World Academy of Sciences and a Member of the South African Young Academy of Science. He has delivered numerous invited talks and keynote presentations globally and holds a patent. Prof. Adebo has been consistently recognized among Stanford University's List of Top 2% Scientists Globally and his research impact is reflected in his ranking as a leading researcher in Food Science in South Africa, across the African continent and globally.

Acknowledgments

The authors appreciate funding received from the University of Johannesburg Research Committee and the Faculty of Science Research Committee, the National Research Foundation (NRF) of South Africa Support for Rated and Unrated Researchers (Grant number: SRUG2204285188) and the South African Medical Research Council (SAMRC) Self-Initiated Research (SIR) Grant. The authors also appreciate the constructive comments of the reviewers for improving the quality of this book.

1

Introduction and Overview

1.1 Introduction

Fermented foods have a rich history in Africa, although their exact origin is challenging to trace due to the absence of a writing culture in most regions. By the medieval ages, Arab travellers had begun to document the presence of fermented foods in Northern and Western Africa. Sour milk was a prominent fermented product, consumed widely in the Guinea Savannah of West and Northern Africa (Farnworth, 2003). An Arab traveller in 1352, Ibn Battuta, noted the prevalence of sour milk in Mali, while another alcoholic drink derived from millet, known as *dolo*, *kimbi* or *merissa*, was mentioned in medieval times (Farnworth, 2003). Brewing in West Africa had become a specialised art by the 8th–16th centuries. Some authors have even opined that fermentation was used as a food preservation technique for cheese since 6000 BC (Figure 1.1) and that the ancient Egyptians from 4000 BC have learnt how to use yeast to preserve bread and wine (Adebo et al., 2023; Paul Ross et al., 2002). Various alcoholic drinks such as *mead* played significant roles in ceremonies, with the ancient Ghana Empire offering them to the deceased (Akyeampong, 1995). The brewing process gradually evolved, with malted millet or sorghum being pounded, cooked, filtered and then fermented (Odunfa & Oyewole, 1998). Sour porridges, a common food preparation practice south of the Sahara, were made by soaking cereal grains to create a savoury flour and to enhance the flavour of the porridge. The tradition of fermenting grains in water for a short time to create a sharp and sweet-tasting drink was noted by German traveller Nachtigal (Odunfa, 1988).

In Sudan, evidence from ancient inscriptions and archaeological sites indicate a long history of fermenting practices. The Greeks noted

DOI: 10.1201/9781003658320-1

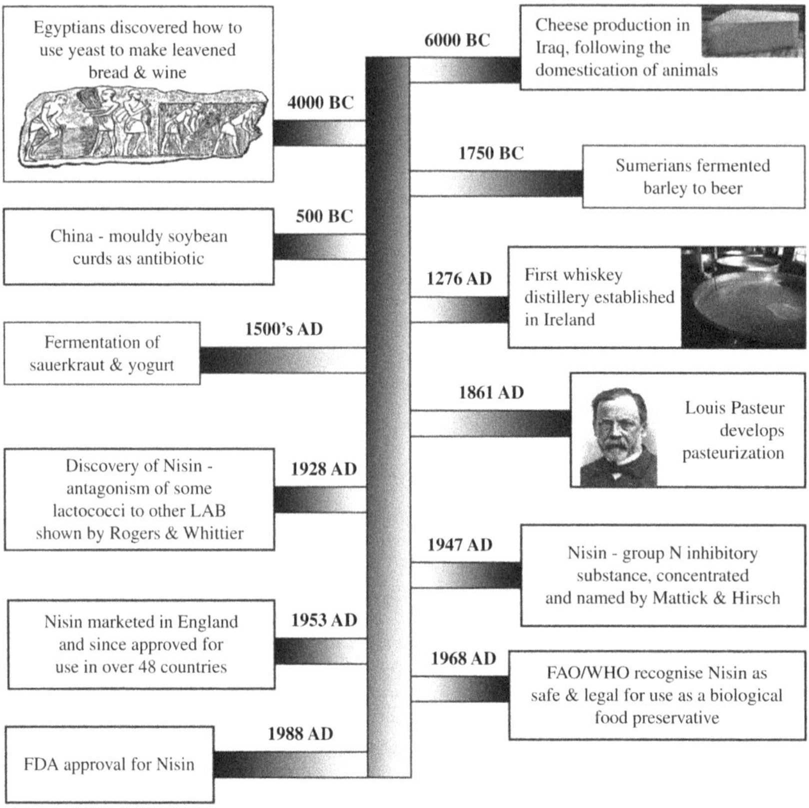

Figure 1.1 Major events in food fermentation and preservation through the years. (From Paul Ross et al., 2002.)

brewing practices in Ethiopia dating back to 7 BC, and the Meroitic civilisation from 690 BC to AD 560 had pottery styles commensurate with wine brewing (Dirar, 1993). Despite the dominance of grape wine in Sudanese history, date wine also persisted, and pottery styles suggest a long association with wine-brewing practices. Overall, fermented foods, including sour milk, alcoholic drinks and porridges, have deep roots in African cultures, shaping dietary practices and culinary traditions over centuries.

According to Panda and Ray (2016), fermentation of food crops can be classified into eight classes, which include fermentation producing textured vegetable protein as meat substitutes, lactic acid fermentation, alcoholic fermentation, acetic acid/vinegar fermentation, alkaline fermentation, leavened breads, flat unleavened breads and high salt sauce and paste fermentation. The microbiology of African fermented foods is a diverse and intricate field encompassing a wide range of traditional food products across the continent.

Fermentation is a traditional food preservation technique that involves the activity of microorganisms, primarily bacteria, yeast and moulds. These microorganisms play crucial roles in shaping the flavour, texture and nutritional content of fermented products (Behera et al., 2019). Fermented foods hold cultural and nutritional significance in various African cultures, contributing to dietary diversity. Examples of African fermented foods include *ogi*, *uji* and *kenkey* in West Africa, injera in East Africa and mahewu in Southern Africa. The microbial communities which play a role in these fermentations are often complex, with lactic acid bacteria (LAB), yeast and *Bacillus* species being commonly identified (Anyogu et al., 2021; Franz et al., 2014; Wafula et al., 2022). Understanding the microbiology of African fermented foods is essential for preserving culinary traditions and exploring the potential health benefits associated with these products.

African fermented foods exhibit a rich diversity, differentiated based on the type of fermentation induced, namely, lactic acid, alkaline or alcoholic fermentation (Figure 1.2) (Obafemi et al., 2022; Zannou et al., 2022). LABs play a central role in lactic acid fermentation. These bacteria convert sugars present in the food matrix into lactic acid, creating an acidic environment that inhibits the growth of spoilage and pathogenic microorganisms. LAB such as *Lactobacillus* and *Pediococcus* species are crucial for shaping the sensory attributes and safety of the final fermented products. This imparts the distinctive sour taste to well-known African fermented products such as *amasi*, *ogi* and *umqombothi*. LABs also contribute to the production of metabolites, including antimicrobial compounds and bioactive molecules, further enhancing the nutritional profile of the fermented foods (Khubber et al., 2022). The microbiology of lactic acid fermentation in African fermented foods is a specialised field within food microbiology. It focuses on the processes and microorganisms involved in the fermentation of foods under conditions that promote lactic acid production. Studying the microbiology of lactic acid fermentation in African foods involves characterising the diversity and functionality of LAB strains, understanding fermentation kinetics and optimising conditions for desirable microbial activity. Molecular techniques such

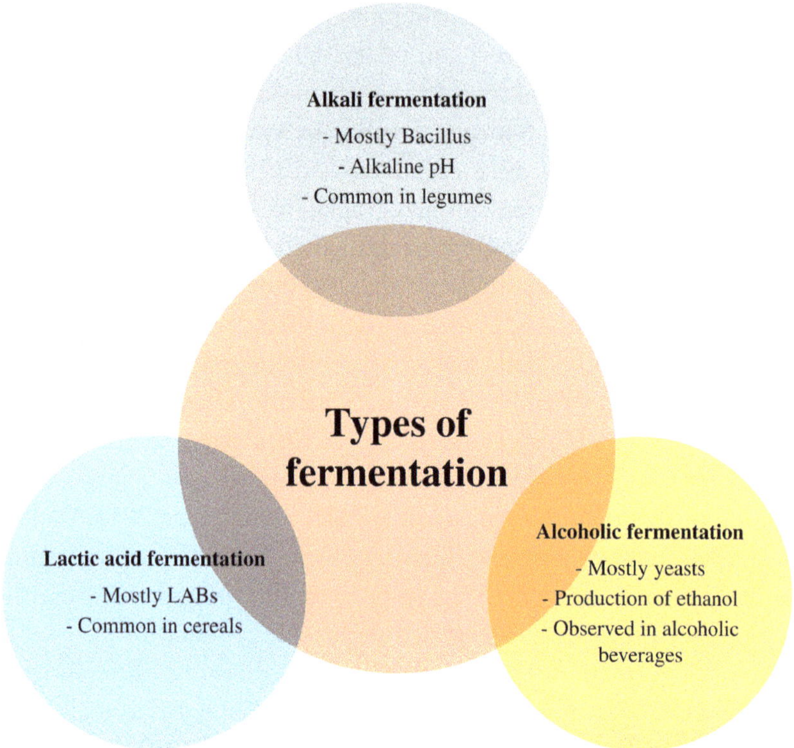

Figure 1.2 Forms of fermentation.

as DNA sequencing and polymerase chain reaction (PCR) are often employed to identify and classify LAB species involved in fermentation. This knowledge not only helps preserve traditional food practices but also contributes to the development of improved and standardised fermentation processes for enhanced food safety and quality. Lactic acid fermentation is a traditional method widely employed in various African communities to process foods, providing distinct flavours and textures and preserving qualities.

Alkaline fermentation, in contrast, requires microorganisms that thrive in alkaline conditions, exemplified in the fermentation of African legumes such as the African locust beans. Initiated by the *Bacillus* species, this process produces alkaline compounds, resulting in condiments like *dawadawa*, *soumbala* and *bikalga* (Compaoré et al., 2013; Jeyaram et al., 2011). The microbiology of alkaline fermentation in African fermented foods focuses on the processes and microorganisms involved in the fermentation of foods under alkaline conditions. Alkaline fermentation is a traditional method employed in various African communities to process certain foods, giving them unique flavours, textures and nutritional profiles. One prominent example is the alkaline fermentation of the Bambara groundnut (*Vigna subterranea*) to produce *dawadawa*, a popular African condiment (Akanni et al., 2018). In alkaline fermentation *Bacillus* species are often key microbial players. These bacteria contribute to the fermentation process, breaking down substrates in the food matrix and producing metabolites that influence the final product's characteristics. The microbiological study of alkaline fermentation investigates the diversity, functionality and dynamics of *Bacillus* species during this process. Techniques such as molecular typing, DNA extraction and sequencing, matrix-assisted laser desorption ionisation-time of flight mass spectrometry (MALDI-TOF MS) identification and sensory analysis are commonly employed to elucidate the microbial composition and factors affecting the fermentation (Akanni et al., 2018; Mbozo et al., 2017; Owusu-Kwarteng et al., 2022). Understanding the microbiology of alkaline fermentation in African foods not only preserves traditional culinary practices but also opens avenues for improving food safety, quality and the nutritional aspects of these fermented products.

Alcoholic fermentation is a common process in the production of African fermented foods and is used in the creation of numerous traditional beverages. This microbiological transformation involves the conversion of sugars into ethanol and carbon dioxide by yeast, typically *Saccharomyces cerevisiae* strains. In the context of African fermented foods, alcoholic fermentation is commonly employed in the production of indigenous beverages such as palm wine, sorghum beer and millet beer (Adebo, 2020; Motlhanka et al., 2018; Sawadogo-Lingani et al., 2021). The process begins with the yeast fermentation of sugars present in the raw materials, such as palm sap or malted grains. The yeast cells metabolise the sugars, producing ethanol and carbon dioxide as byproducts. The microbiota involved in alcoholic fermentation also includes LAB, which can contribute to the flavour profile of the final product. The specific strains of yeast and bacteria, as well as the fermentation conditions, play crucial roles in determining the sensory characteristics and quality of the alcoholic beverages. Overall, the microbiology of alcoholic fermentation in African fermented foods showcases a dynamic interplay of microorganisms that significantly influence the taste, aroma and traditional significance of these culturally important beverages.

Lactic acid, alkaline and alcoholic fermentations are deeply ingrained in African culinary traditions, contributing to flavour development but also to the preservation and enhancement of nutritional properties in the African fermented foods. The choice of fermentation method often reflects African communities' cultural practices and regional preferences. All these fermentation methods are deeply embedded in African culinary traditions, contributing to the flavour complexity of the foods and playing a key role in preserving and enhancing the nutritional qualities of these cultures.

1.2 Importance of Microorganisms in African Fermented Foods

Microorganisms play a crucial role in the food system of African fermented foods, contributing to the development, safety and nutritional characteristics of these traditional products. The importance of microorganisms in African fermented foods can be outlined in several key statements. In the fermentation process, microorganisms such as LABs, *Bacillus* and yeasts drive the fermentation process in African foods. This process transforms raw materials into the final fermented products, providing unique flavours, textures and aromas. Fermentation acts as a natural preservation method. The organic acids and other metabolic byproducts produced by microorganisms create an inhospitable environment for spoilage and pathogenic bacteria, extending the shelf life of the fermented foods.

Microorganisms contribute to the nutritional quality of African fermented foods. They can increase the bioavailability of specific nutrients, produce vitamins and break down complex compounds into more digestible forms (Adebo, 2020; Adebo et al., 2022). The microorganisms involved in fermentation often have cultural and traditional significance. They are part of the heritage and identity of communities, influencing local dietary habits and culinary practices. Different microorganisms contribute to the diversity of fermented foods across Africa. Through the use of specific strains of LAB, *Bacillus*, moulds or other microbes, each community's selection of microorganisms results in a wide array of unique fermented products.

Certain fermented foods also contain probiotic microorganisms that can confer health benefits to consumers. Probiotics contribute to gut health and may enhance the immune system. The sensory characteristics of fermented foods, including taste, smell and texture, are influenced by the metabolic activities of microorganisms. This contributes to the distinctiveness of each fermented product.

1.3 Role of Microorganisms during Fermentations of African Fermented Foods

Microorganisms in African fermented foods, including bacteria (LAB and *Bacillus* species), yeasts and moulds, serve as the primary agents responsible for the fermentation process. They transform raw materials into the final fermented products through various metabolic activities. LABs are commonly involved in African fermentations and work by producing organic acids (e.g., lactic acid) as metabolic byproducts (Abiola et al., 2022). These acids contribute to the sour taste of fermented foods and act as natural preservatives, inhibiting the growth of spoilage and pathogenic microorganisms. Microorganisms generate a wide range of volatile compounds during fermentation, influencing the aroma and flavour of the final product (Akanni et al., 2018; Compaoré et al., 2013). These compounds contribute to African fermented foods' unique and characteristic sensory profiles. Microorganisms play a crucial role in breaking down complex carbohydrates, proteins and fats into simpler, more digestible forms. This process enhances the nutrients' bioavailability and increases the nutritional value of fermented foods. The acidic environment created by the metabolic activities of microorganisms prevents the growth of spoilage and pathogenic bacteria, extending the shelf life of the fermented products. Likewise, is the role of these microorganisms in the degradation and elimination of antinutrients, heavy metals and toxic components in food (Adebiyi et al., 2019; Adebo et al., 2025; Ma, 2024).

Selecting specific microorganisms for fermentation is often rooted in cultural traditions and practices. These microorganisms contribute to the uniqueness of local fermented foods and are integral to the culinary heritage of communities. Certain fermented foods contain probiotic microorganisms that can confer health benefits. Probiotics contribute to gut health and aid overall digestion and the immune system (Marco et al., 2017). Different microorganisms contribute to the diversity of fermented foods across Africa. The specific strains used fermentation conditions, and regional variations result in various unique products with distinct characteristics. Understanding the role of microorganisms in African fermented foods is essential for preserving cultural traditions, ensuring food safety and exploring opportunities for innovation within local food systems.

References

Abiola, R. R., Okoro, E. K., & Sokunbi, O. (2022). Lactic acid bacteria and the food industry–A comprehensive review. *International Journal of Health Sciences and Research*, *12*(5), 128–142.

Adebiyi, J. A., Kayitesi, E., Adebo, O. A., Changwa, R., & Njobeh, P. B. (2019). Food fermentation and mycotoxin detoxification: An African perspective. *Food Control*, *106*, 106731.

Adebo, J. A., Njobeh, P. B., Gbashi, S., Oyedeji, A. B., Ogundele, O. M., Oyeyinka, S. A., & Adebo, O. A. (2022). Fermentation of cereals and legumes: Impact on nutritional constituents and nutrient bioavailability. *Fermentation*, *8*(2), 63.

Adebo, O. A. (2020). African sorghum-based fermented foods: Past, current and future prospects. *Nutrients*, *12*(4), 1111.

Adebo, O. A., Chinma, C. E., Obadina, A. O., Soares, A. G., Panda, S. K., & Gan, R.-Y. (2023). An insight into indigenous fermented foods for the tropics. In O. A. Adebo, C. E. Chinma, O. Obadina, S. Panda, A. Soares, & R. Y. Gan (Eds.), *Indigenous fermented foods in the tropics* (pp. 1–12). Elsevier.

Adebo, O. A., Gieng, J., & Feng, X. (2025). Editorial: Enhancing nutrient profile, safety, and sustainability with fermentation technology. *Frontiers in Nutrition*, *12*, 1571781.

Akanni, G. B., Naudé, Y., de Kock, H. L., & Buys, E. M. (2018). Diversity and functionality of *Bacillus* species associated with alkaline fermentation of bambara groundnut (*Vigna subterranean* L. *Verdc*) into dawadawa-type African condiment. *European Food Research and Technology*, *244*, 1147–1158.

Akyeampong, E. (1995). Alcoholism in Ghana—A socio-cultural exploration. *Culture, Medicine and Psychiatry*, *19*, 261–280.

Anyogu, A., Olukorede, A., Anumudu, C., Onyeaka, H., Areo, E., Adewale, O., Odimba, J. N., & Nwaiwu, O. (2021). Microorganisms and food safety risks associated with indigenous fermented foods from Africa. *Food Control*, *129*, 108227.

Behera, S. S., Ray, R. C., Das, U., Panda, S. K., & Saranraj, P. (2019). Microorganisms in fermentation. In *Essentials in fermentation technology*, (pp. 1–39). Springer International Publishing.

Compaoré, C. S., Nielsen, D. S., Ouoba, L. I., Berner, T. S., Nielsen, K. F., Sawadogo-Lingani, H., Diawara, B., Ouédraogo, G. A., Jakobsen, M., & Thorsen, L. (2013). Co-production of surfactin and a novel bacteriocin by *Bacillus subtilis* subsp. *subtilis* H4 isolated from *Bikalga*, an African alkaline *Hibiscus sabdariffa* seed fermented condiment. *International Journal of Food Microbiology, 162*(3), 297–307.

Dirar, H. A. (1993). *The indigenous fermented foods of the Sudan. A study in African food and nutrition*. CAB International Wallingford.

Farnworth, E. R. (2003). *Handbook of fermented functional foods*. CRC Press.

Franz, C. M., Huch, M., Mathara, J. M., Abriouel, H., Benomar, N., Reid, G., Galvez, A., & Holzapfel, W. H. (2014). African fermented foods and probiotics. *International Journal of Food Microbiology, 190*, 84–96.

Jeyaram, K., Romi, W., Singh, T. A., Adewumi, G. A., Basanti, K., & Oguntoyinbo, F. A. (2011). Distinct differentiation of closely related species of *Bacillus subtilis* group with industrial importance. *Journal of Microbiological Methods, 87*(2), 161–164.

Khubber, S., Marti-Quijal, F. J., Tomasevic, I., Remize, F., & Barba, F. J. (2022). Lactic acid fermentation as a useful strategy to recover antimicrobial and antioxidant compounds from food and by-products. *Current Opinion in Food Science, 43*, 189–198.

Ma, X. (2024). Heavy metals remediation through lactic acid bacteria: Current status and future prospects. *Science of the Total Environment, 946*, 174455.

Marco, M. L., Heeney, D., Binda, S., Cifelli, C. J., Cotter, P. D., Foligné, B., Gänzle, M., Kort, R., Pasin, G., Pihlanto, A., & Smid, E. J. (2017). Health benefits of fermented foods: Microbiota and beyond. *Current Opinion in Biotechnology, 44*, 94–102.

Mbozo, A. B. V., Kobawila, S. C., Anyogu, A., Awamaria, B., Louembe, D., Sutherland, J. P., & Ouoba, L. I. I. (2017). Investigation of the diversity and safety of the predominant *Bacillus pumilus sensu lato* and other *Bacillus* species involved in the alkaline fermentation of cassava leaves for the production of Ntoba Mbodi. *Food Control, 82*, 154–162.

Motlhanka, K., Zhou, N., & Lebani, K. (2018). Microbial and chemical diversity of traditional non-cereal based alcoholic beverages of Sub-Saharan Africa. *Beverages, 4*(2), 36.

Obafemi, Y. D., Oranusi, S. U., Ajanaku, K. O., Akinduti, P. A., Leech, J., & Cotter, P. D. (2022). African fermented foods: Overview, emerging benefits, and novel approaches to microbiome profiling. *NPJ Science of Food, 6*(1), 15.

Odunfa, S. A. (1988). African fermented foods: From art to science. *MIRCEN Journal of Applied Microbiology and Biotechnology, 4*, 259–273.

Odunfa, S. A., & Oyewole, O. B. (1998). African fermented foods. In B. J. W. Wood (Ed.), *Microbiology of fermented foods* (2nd ed., Vol. 2, pp. 713–752). Blackie Academic and Professional.

Owusu-Kwarteng, J., Parkouda, C., Adewumi, G. A., Ouoba, L. I. I., & Jespersen, L. (2022). Technologically relevant *Bacillus* species and microbial safety of West African traditional alkaline fermented seed condiments. *Critical Reviews in Food Science and Nutrition, 62*(4), 871–888.

Panda, S. K., & Ray, R. C. (2016). Fermented foods and beverages from tropical roots and tubers. In *Tropical roots and tubers: Production, processing and technology* (pp. 225–252). Wiley.

Paul Ross, R., Morgan, S., & Hill, C. (2002). Preservation and fermentation: Past, present and future. *International Journal of Food Microbiology, 79*, 3–16.

Sawadogo-Lingani, H., Owusu-Kwarteng, J., Glover, R., Diawara, B., Jakobsen, M., & Jespersen, L. (2021). Sustainable production of African traditional beers with focus on dolo, a West African sorghum-based alcoholic beverage. *Frontiers in Sustainable Food Systems*, *5*, 672410.

Wafula, E. N., Muhonja, C. N., Kuja, J. O., Owaga, E. E., Makonde, H. M., Mathara, J. M., & Kimani, V. W. (2022). Lactic acid bacteria from African fermented cereal-based products: Potential biological control agents for mycotoxins in Kenya. *Journal of Toxicology, 2022.*

Zannou, O., Agossou, D. J., Miassi, Y., Agani, O. B., Aisso, M. D., Chabi, I. B., Kpoclou, Y. E., Azokpota, P., & Koca, I. (2022). Traditional fermented foods and beverages: Indigenous practices of food processing in Benin Republic. *International Journal of Gastronomy and Food Science, 27*, 100450.

2

Microbiology of African Fermented Cereals

2.1 Introduction

Indigenous African cereal grains encompass a diverse array of nutrient-rich staples that have sustained communities across the continent for centuries. These grains have played a central role in African diets and cultural practices. Beyond their culinary significance indigenous cereal grains possess unique nutritional profiles, are resilient to harsh environmental conditions and serve as suitable substrates for fermentation, making them essential components of food security and agricultural sustainability in Africa. A plethora of African fermented products are made from African cereals (Table 2.1).

Table 2.1 Most Common Indigenous Fermented Foods and Beverages from Africa

PRODUCT	RAW MATERIAL	PRODUCT FORM	COUNTRY/REGION
Aceda	Sorghum	Porridge	Sudan
Balady	Wheat	Flatbread	Egypt
Ben-saalga	Pearl millet	Weaning food	Burkina Faso, Ghana
Bouza	Wheat	Gruel	Egypt
Burukutu	Maize, millet, sorghum	Beverage	Ghana, Nigeria
Bushera	Sorghum, millet	Beverage	Uganda
Busaa/Bussa	Maize	Beverage	East Africa, Kenya
Chibuku	Sorghum	Beverage	Zimbabwe
Dolo	Sorghum	Beverage	Burkina Faso, Togo
Enturire	Sorghum	Beverage	Uganda
Gowé	Sorghum	Porridge	Benin
Humulur	Sorghum	Gruel	Sudan
Hussuwa	Sorghum	Porridge	Sudan
Ikigage	Sorghum	Beverage	Rwanda

(Continued)

DOI: 10.1201/9781003658320-2

Table 2.1 Most Common Indigenous Fermented Foods and Beverages from Africa *(Continued)*

PRODUCT	RAW MATERIAL	PRODUCT FORM	COUNTRY/REGION
Injera	Teff, sorghum, maize, finger millet, wheat, or barley	Sourdough, bread	Ethiopia
Kachasu	Maize	Beverage	Zimbabwe
Kenke/Kenkey	Maize	Dumpling	Ghana
Koko	Millet	Gruel	Ghana
Kisra	Sorghum, pearl millet	Pancake, flatbread, sourdough	Sudan
Kunun-zaki	Millet, sorghum	Beverage	Nigeria
Lting/Joala	Sorghum	Beverage	South Africa
Mahewu	Maize	Gruel	South Africa
Mangisi	Millet	Beverage	Zimbabwe
Mbege	Millet	Beverage	Tanzania
Merissa	Sorghum, millet	Beverage	Sudan
Ogi	Maize, sorghum, millet	Gruel	West Africa
Ori-ese	Sorghum	Porridge	Nigeria
Orubisi	Maize, millet	Beverage	Tanzania
Otika	Maize	Beverage	Nigeria
Pito	Maize, sorghum	Beverage	Nigeria
Poto poto	Maize	Gruel	Congo
Tchapalo	Sorghum	Beverage	Ivory Coast
Tella	Sorghum	Beverage	Ethiopia
Tchoukoutou	Sorghum	Beverage	Benin
Thobwa	Maize, millet, sorghum	Beverage	Malawi
Ting	Sorghum, maize	Porridge, sourdough	Southern Africa
Uji	Maize, sorghum, millet	Porridge	Kenya, Tanzania, Uganda
Umqombothi	Maize, sorghum	Beverage	Southern Africa

Source: Adapted from Adebo et al. (2023).

The microbiology of African fermented cereals is a complex and dynamic process that involves the interaction of various microorganisms to transform raw cereals into palatable and nutritious products. Lactic acid bacteria (LAB), particularly species such as *Lactobacillus*, *Pediococcus* and *Leuconostoc*, are key players in this fermentation process (Kayitesi et al., 2023; Soro-Yao et al., 2014). These bacteria utilise sugars present in cereals to produce lactic acid, leading to a decrease in pH that contributes to the characteristic sour taste of fermented cereals (Soro-Yao et al., 2014). LAB also play a crucial role in inhibiting the growth of harmful microorganisms through competition for nutrients and the production of antimicrobial compounds. Species like *Lactobacillus* play a central role in the fermentation process, contributing to nutritional value, organoleptic properties, shelf life and food safety (Kayitesi et al., 2023). LAB dominance in naturally occurring microbial populations assists in generating fermentable sugars used as energy sources. LAB produces antibiotics, hydrogen peroxide and organic acids, imparting antibiosis and inhibiting the growth of certain microorganisms (Afolayan et al., 2017; Chelule et al., 2010). African cereal-based fermented products contain a diverse mixture of microorganisms due to spontaneous fermentation, which involves several stages of lactic acid fermentation followed by alcoholic fermentation. Predominant LAB species include *Lactobacillus acidophilus* and *Lactiplantibacillus plantarum*.

In addition to LABs, yeasts and moulds are often involved in the fermentation of African cereals. Yeasts, such as *Saccharomyces* and *Candida*, contribute to the production of ethanol and carbon dioxide, influencing the flavour and texture of the final product (Mashau et al., 2021). Moulds, including *Aspergillus* and *Rhizopus*, produce enzymes that break down complex carbohydrates and proteins into simpler compounds, enhancing the nutritional profile of fermented cereals.

African people have been producing fermented products, particularly cereal-based food (maize, sorghum, millet), since 3500 BC (Adebo, 2020; Diaz et al., 2019; Konfo et al., 2015). Using cereals as a fermentation substrate is an effective way to develop functional foods as they are rich in nutrients that can be easily assimilated by probiotics (Ignat et al., 2020). The specific composition of microbial communities and their activities can vary depending on factors such as the type of cereal, where it is made and traditional fermentation practices. Understanding the microbiology of African fermented cereals is essential not only for the production of culturally significant foods but also for optimising the nutritional content, safety and sensory characteristics of these products. Maize (*Zea mays*) and sorghum are the most commonly used substrates in African cereal-based fermented products, either singly or in combination with other cereal grains such as millet and teff, as depicted in Table 2.2. This chapter will focus on African cereal-based fermented products made from maize or sorghum across various regions in Africa.

Table 2.2 Major Fermenting Microbial Species in Common Indigenous Fermented Foods and Beverages Across Africa

PRODUCT	COUNTRY/REGION	PRODUCT FORM	SUBSTRATE	PREDOMINANT MICROORGANISMS IN FERMENTATION	REFERENCE
Agidi	Benin, Nigeria	Porridge	Maize	*Lactiplantibacillus plantarum, Lb. acidophilus, Leuconostoc mesenteroides, Pediococcus* spp., *Saccharomyces cerevisiae, Candida krusei*	Adegbehingbe et al. (2019)
Balady	Egypt	Flatbread	Wheat	*Saccharomyces cerevisiae, Lactiplantibacillus plantarum*	Mohsen et al. (2016)
Ben-Saalga	Burkina Faso, Ghana	Weaning food	Pearl millet	*Limosilactobacillus fermentum, Lp. paraplantarum, Li. salivarius, Lb. delbrueckii, Lb. amylolyticus, Lb. reuteri, Lactococcus lactis, Leuconostoc mesenteroides, Pediococcus acidilactici, P. pentosaceus, Streptococcus gallolyticus, Weissella confusa*	Turpin et al. (2013), Soro-Yao et al. (2014)
Burukutu	Ghana, Nigeria	Beverage	Maize, millet, sorghum	*Saccharomyces cerevisiae, Levilactobacillus brevis, Lb. lactis, Lp. plantarum, Lm. fermentum, Lb. acidophilus*	Ire et al. (2020)
Busaa	Kenya	Beverage	Maize	*Lactobacillus helveticus, Ligilactobacillus salivarius, Pediococcus damnosus, Pediococcus parvulus, Candida krusei, Saccharomyces cerevisiae*	Nout (1980)

Bushera	Uganda	Beverage	Sorghum, millet	*Streptococcus* spp., *Weissella* spp., *Lactobacillus* spp., *Leuconostoc* spp., *Lactococcus* spp., *Pichia* spp., *S. cerevisiae*, *C. krusei*	Mukisa et al. (2012)
Dolo	Burkina Faso, Togo	Beverage	Sorghum	*Limosilactobacillus fermentum*, *Lb. delbrueckii*, *P. acidilactici*, *S. cerevisiae*, *Candida kefyr*, *C. krusei*, *C. lusitaniae*, *C. albicans*, *C. tropicalis*	Sawadogo-Lingani et al. (2007), Sanata et al. (2017)
Gowé	Benin	Porridge	Sorghum	*Lactobacillus mucosae*, *Lm. fermentum*, *Weissella kimchii*, *W. confusa*, *Pediococcus acidilactici*, *P. pentosaceus*, *Clavispora lusitaniae*, *Candida krusei*, *C. tropicalis*, *Kluyveromyces marxianus*, *Pichia anomala*	Vieira-Dalodé et al. (2007), Greppi et al. (2013)
Hussuwa	Sudan	Porridge	Sorghum	*Limosilactobacillus fermentum*, *P. acidilactici*, *Enterobacter faecium*, *Lb. saccharolyticum*, *Gluconobacter oxydans*, *Acetobacter xylinum*, *Saccharomyces cerevisiae*	El Sheikha and Montet (2014), Franz and Holzapfel (2019)
Ikigage	Rwanda	Beverage	Sorghum	*Saccharomyces cerevisiae*, *Candida inconspicua*, *C. magnolia*, *C. humilis*, *Issatchenkia orientalis*, *Lactobacillus buchneri*, *Lm. fermentum*	Lyumugabe et al. (2010)

(Continued)

Table 2.2 Major Fermenting Microbial Species in Common Indigenous Fermented Foods and Beverages Across Africa *(Continued)*

PRODUCT	COUNTRY/REGION	PRODUCT FORM	SUBSTRATE	PREDOMINANT MICROORGANISMS IN FERMENTATION	REFERENCE
Injera	Ethiopia	Sourdough, bread	Teff, sorghum, maize, finger millet, wheat, or barley	*Saccharomyces cerevisiae, Kazachstania humilis, Pichia fermentans, P. norvegensis, P. occidentalis, Candida humilis, C. tropicalis, Saccharomyces exiguous, Kazachstania bulderi, Enterococcus lactis, Lactocaseibacillus paracasei, Lactobacillus species, Pediococcus pentosaceus, P. acidilactici, Limosilactobacillus fermentum, Lp. plantarum, Lactococcus piscium, Lb. raffinolactis, Leuconostoc mesenteroides* subsp. *dextranicum, Lc. mesenteroides, Enterococcus casseliflavus*	Tadesse et al. (2019), Neela and Fanta (2020), Dessalegn and Andualem (2023a), Dessalegn and Andualem (2023b)
Kenke/Kenkey	Ghana	Dumpling	Maize	*Limosilactobacillus fermentum, Lv. brevis, Lp. plantarum, Pediococcus pentosaceus, P. acidilactici, Candida krusei, Saccharomyces cerevisiae*	Annan (2013), El Sheikha and Montet (2014)
Koko	Ghana	Gruel	Millet	*Lm. fermentum, Li. salivarius, Lp. paraplantarum, Weissella confusa, Pediococcus pentosaceus, P. acidilactici*	Lei and Jakobsen (2004)
Kisra	Sudan	Pancake, flat bread, sourdough	Sorghum, pearl millet	*Lm. fermentum, Lb. reuteri, Lb. confusus, Lv. brevis, Pediococcus pentosaceus, Enterococcus faecium*	Eltayeb et al. (2020)

Kunun-zaki	Nigeria	Beverage	Millet, sorghum	*Limosilactobacillus fermentum, Lp. plantarum, Lb. leichmannii, Lb. lactis, S. cerevisiae, Streptococcus faecium, Micrococcus acidophilus*	Efiuvwevwere and Akona (1995), Amusa and Ashaye (2009)
Lemzeiet	Algeria	Couscous	Wheat	*Enterococcus faecium, E. faecalis, E. hirae, Levilactobacillus brevis, Lb. buchneri, Lb. namurensis, Lb. rapi, Lb. coryniformis, Lb. curvatus, Lb. pentosus, Leuconostoc mesenteroides, Leu. fallax, Pediococcus pentosaceus, P. acidilactici, Streptococcus mitis, Weissella paramesenteroides*	Merabti et al. (2015)
Mahewu	Southern Africa	Gruel	Maize	*Lactococcus lactis* subsp. *lactis, Lacticaseibacillus casei, Lm. fermentum, Lp. plantarum, Lb. rossiae, Lb. lactis, Leuconostoc holzapfelii, Pediococcus pentosaceus, Weissella cibaria, W. confusa, Saccharomyces pombe, S. cerevisiae, Candida glabrata, C. krusei*	Idowu et al. (2016), Fadahunsi and Soremekun (2017), Pswarayi and Gänzle (2019)
Mawe	Benin	Gruel	Maize	*Limosilactobacillus fermentum, Lv. brevis, Lb. reuteri, Pediococcus acidilactici, Weissella confusa, Pichia kudriavzevii, Kluyveromyces marxianus, Saccharomyces cerevisiae*	Houngbédji et al. (2018)

(Continued)

Table 2.2 Major Fermenting Microbial Species in Common Indigenous Fermented Foods and Beverages Across Africa *(Continued))*

PRODUCT	COUNTRY/REGION	PRODUCT FORM	SUBSTRATE	PREDOMINANT MICROORGANISMS IN FERMENTATION	REFERENCE
Ogi	West Africa	Gruel	Maize, sorghum, millet	*Weissella paramesenteroides, Lactobacillus delbrueckii* subsp. *bulgaricus, Lactobacillus cellobiosus, Lb. vaccinostercus, Lp. plantarum, Lb. acidophilus, Lv. brevis, Lb. rossiae, Lm. fermentum, Lb. helveticus, Lb. pantheris, Lb. bifermentans, Lb. nantensis, Leuconostoc lactis, Leuc. Paramesenteroides, Acetobacter pasteurianus, Paucilactobacillus vaccinostercus, Companilactobacillus nantensis, Saccharomyces cerevisiae, Candida krusei, C. tropicalis, Geotrichum candidum, G. fermentans, Rhodotorula graminis*	Ampe and Miambi (2000), Omemu et al. (2007), Oguntoyinbo et al. (2011), Oguntoyinbo and Narbad (2012), Oyedeji et al. (2013), Greppi et al. (2013), Afolayan et al. (2017)
Pito	West Africa	Beverage	Maize, sorghum	*Limosilactobacillus fermentum, Lp. plantarum, Lb. helveticus, La. paracasei* and *Lv. brevis. Lb. delbrueckii, Saccharomyces cerevisiae, Debaryomyces hansenii, Pichia anomala*	Sawadogo-Lingani et al. (2007), Zaukuu et al. (2016), Tokpohozin et al. (2017), Djameh et al. (2019)
Poto poto	Congo	Gruel	Maize	*Lactobacillus gasseri, Lp. plantarum/ paraplantarum, Lb. acidophilus, Lb. delbrueckii, Lb. reuteri, La. casei*	Abriouel et al. (2006)

Tchapalo	Ivory Coast	Beverage	Sorghum	*Pediococcus acidilactici*, *P. pentosaceus*, *Limosilactobacillus fermentum*	Tano et al. (2020)
Tella	Ethiopia	Beverage	Sorghum	*Acetobacter xylinum*, *Lactobacillus pasteurianum*, *Saccharomyces cerevisiae*, *Saccharomyces carlsbergensis*	Tekle et al. (2019)
Ting	Southern Africa	Porridge, sourdough	Sorghum, maize	*Lacticaseibacillus casei*, *Lb. coryniformis*, *Lb. curvatus*, *Lm. fermentum*, *Lb. harbinensis*, *Lb. parabuchneri*, *Lp. plantarum*, *Lb. reuteri*, *La. rhamnosus*	Adebo (2020)
Uji	Kenya, Tanzania, Uganda	Porridge	Maize, sorghum, millet	*Lactiplantibacillus plantarum*, *Lb. fermenti*, *Lb. buchneri*, *Lb. cellobiosus*, *Pediococcus acidilactici*, *P. pentosaceus*	Onyango et al. (2000)
Umqombothi	Southern Africa	Beverage	Maize, sorghum	*Ligilactobacillus*, *Lactiplantibacillus*, *Lactobacillus*, *Lactococcus*, *Kosakonia cowanii*, *Apilactobacillus pseudoficulneus*, *Vibrio alginolyticus*, *Apiotrichum laibachii*, *P. anomala*, *P. fermentans*, *Endomycopsis fibuligera*, *Kloeckera apiculata*, *S. cerevisiae*, *K. marxianus*, *C. krusei* and *C. tropicalis*	Hlangwani et al. (2020), Hlangwani et al. (2023)

Abbreviations: La., Lacticaseibacillus; Lb., Lactobacillus; Li., Ligilactobacillus; Lm., Limosilactobacillus; Lp., Lactiplantibacillus; Lv., Levilactobacillus.

2.2 Notable Fermented Cereal-Based Products from West Africa

2.2.1 *Ogi*

Ogi, also known as pap, *akamu* or *koko*, is a traditional fermented cereal gruel commonly used as a non-alcoholic starchy weaning food. It is usually produced from the fermentation of maize or other cereals by a traditional fermentation method (Ekpa et al., 2019; Mashau et al., 2021). The fermentation of *ogi* involves a simple yet crucial process that transforms maize kernels or flour into a nutritious and digestible food product. *Ogi* is a sour, fine paste beverage that, when cooked, produces a thin semi-solid porridge (Nwosu & Oyeka, 1998). *Ogi* has a smooth texture and a sour taste resembling that of yoghurt. Production is usually homemade, through a spontaneous fermentation process done in households.

The main ingredient of *ogi* is maize (corn) kernels or flour, although other grains such as millet or sorghum may also be used. The maize is thoroughly cleaned to remove any dirt or debris and then dried under the sun to reduce moisture content. Once dried, the maize is milled into a fine powder to obtain maize flour, which serves as the base for *ogi*. The maize flour is mixed with water in a ratio that typically ranges from 1:2 to 1:4 (maize flour to water). The mixture is stirred vigorously to ensure even distribution of the flour and water, resulting in a smooth, thick paste-like consistency (Oyarekua & Eleyinmi, 2004). The mixed maize paste is left to ferment for a period of one to three days, depending on the desired acidity and flavour of the *ogi* (Figure 2.1).

The *ogi* fermentation is a two-phase process. In the first phase, naturally occurring microorganisms such as Gram-positive bacteria and yeast present in the environment or introduced from previous batches of *ogi* initiate the fermentation process (Anumudu et al., 2018). These microorganisms metabolise the complex carbohydrates present in the maize flour into fermentable sugars. The second phase is dominated by LAB species and yeasts, producing lactic acid and other organic acids, as well as carbon dioxide and alcohol as byproducts (Anumudu et al., 2018; Nwosu & Oyeka, 1998; Oyedeji et al., 2013). The production of lactic acid lowers the pH of the mixture, creating a slightly acidic environment that inhibits the growth of harmful bacteria while promoting the growth of beneficial microorganisms. As fermentation progresses, the texture of the mixture changes from thick and starchy

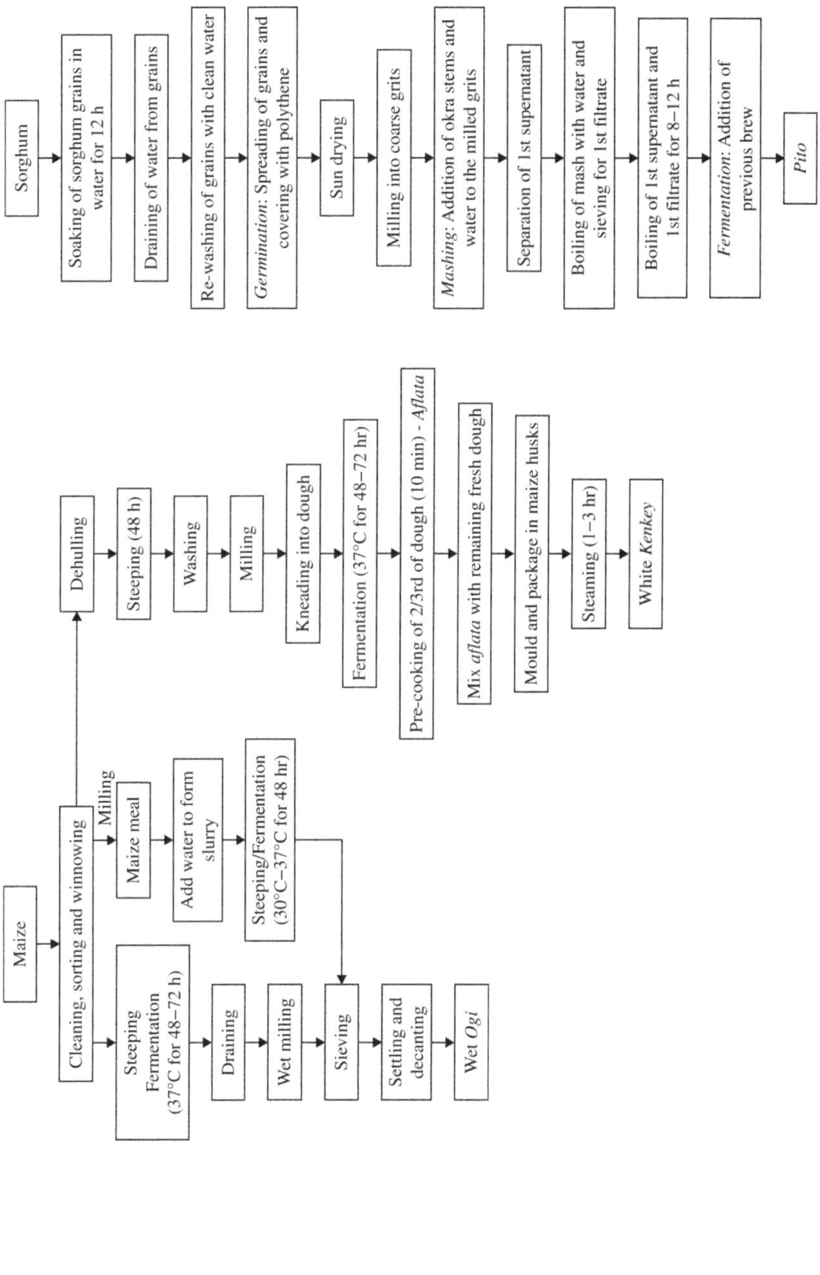

Figure 2.1 Flow chart of the production of some maize- or sorghum-based African fermented food products from West Africa.

to smooth and slightly sour, with a characteristic tangy flavour. The predominant microorganism present has been reported to be LAB; however, consortia of other microbial flora have also been identified throughout the fermentation process.

The steep liquor of maize flour and other cereals generally has a pH of 6.4 on the first day, which later reduces to 3.4 at the end of fermentation, usually around three to five days. The bacterial count after fermentation of the steep liquor can be as high as 9 log CFU/g, while the fungal count is 5 log CFU/g (Anumudu et al., 2018). Bacteria isolated include *Lactobacillus* species, *Bacillus* species, *Corynebacterium* species, *Streptococcus* species and *Clostridium* species (Oyedeji et al., 2013). The isolated fungal genera include *Aspergillus* species, *Fusarium* species, *Penicillium* species, *Saccharomyces* species and *Candida* species. A succession pattern of microorganisms involved in the fermentation process can be observed. Various LAB and yeast species tend to be the predominant microorganisms by the end of fermentation (Table 2.2). The microbial succession pattern in *ogi* fermentation impacts the unique organoleptic qualities and fermentation properties of the *ogi*, with the LAB producing acids and inhibiting other bacteria growth, and yeast contributing to the flavour development of the product.

2.2.2 Kenkey

Kenkey is a prominent traditional fermented food consumed in Ghana, West Africa. It is made primarily from fermented maize or corn dough and is part of the staple diet in the area. The process of preparing this dish begins with soaking whole maize grains in water for several hours, usually overnight (Halm et al., 2004). This soaking helps to soften the grains and initiate the fermentation process. After soaking, the maize is thoroughly washed to remove any impurities. The soaked corn is ground into flour, which is then mixed with warm water to start a fermentation process that spans approximately two to three days until a dough is formed (Figure 2.1). During fermentation, naturally occurring microorganisms such as LAB (*Lactobacillus* species, *Pediococcus acidilactici*, *Leuconostoc mesenteroides* and *Leuc. fermentum*) and yeasts ferment the maize dough (Table 2.2). These microorganisms metabolise the carbohydrates in the maize, producing lactic acid, other organic acids, carbon dioxide and ethanol (Annan, 2013; El Sheikha & Montet, 2014). The fermentation process contributes to *kenkey*'s characteristic sour flavour and tangy aroma. Following fermentation, the dough is divided into two equal portions. One portion undergoes partial cooking in a large pot of water for about 10 minutes, with constant and vigorous stirring, resulting in what is termed *aflata*. This cooked half of the dough is then combined with the remaining uncooked portion and thoroughly mixed. The mixture is divided into individual serving-sized pieces and tightly wrapped in banana leaves, cornhusks or foil. These wrapped dough packets are placed on a wire rack above water in a large pot and boiled/steamed for 1–3 hours. The end product, *kenkey*, is typically served alongside a sauce or any fish or meat dish. There are several variants of *kenkey*, such as *Ga* and *Fante* (Nyanzi & Jooste, 2012).

2.2.3 Pito

Pito is a traditional African alcoholic beverage commonly consumed in Ghana, Nigeria and other West African countries. *Pito* is produced from local grains, mainly sorghum, guinea corn (*Sorghum bicolour*) and other grains such as maize (*Zea mays*) and millet (*Pennisetum glaucum*) (Djameh et al., 2015; Zaukuu et al., 2016). These grains are cleaned, washed and soaked in water for two days to initiate germination. After soaking, the grains are spread out to germinate, usually for two to five days, in baskets lined with moistened banana leaves, with the time depending on the desired characteristics of the final product. Once germination is complete, the grains are dried and roasted to stop further enzymatic activity and to develop flavour compounds. The roasted grains are then milled or crushed to form a coarse powder known as malt (Figure 2.1). The malt is mixed with hot water in a process called mashing, so that the enzymes in the malt can break down starches into fermentable sugars. This step also extracts flavours and colour from the malt (Djameh et al., 2019).

The mashed mixture, known as mash, is transferred to a fermentation vessel or a large pot. A starter culture from a previous batch may be introduced to the mash, which is then left to ferment. Wild yeast and bacteria in the environment also contribute to *pito* fermentation (Djameh et al., 2019; Sawadogo-Lingani et al., 2007;

Tokpohozin et al., 2017). The fermentation vessel is covered or sealed to prevent contamination and allow fermentation to proceed anaerobically (without oxygen). Fermentation typically takes several days to a week. During this time the sugars in the mash are converted into alcohol and carbon dioxide by the action of yeast and other microorganisms (Table 2.2). This process produces *pito*'s characteristic tangy flavour and alcoholic content. After fermentation is complete, the fermented mash is strained to remove solid particles, resulting in a liquid known as "wet" *pito*. Wet *pito* may be consumed as is or it may undergo further maturation or conditioning to develop its flavour. Maturation can occur in the fermentation vessel or in separate containers and may take several days to weeks, depending on the desired taste and aroma profile (Zaukuu et al., 2016). *Pito*, the final product, is a dark brown liquid with a flavour profile ranging from sweet to bitter. It contains lactic acid, sugars and amino acids, and has an alcohol content of approximately 3%. *Geotrichum candidum* and *Lactobacillus* spp. are responsible for the souring, while *Saccharomyces cerevisiae*, *Debaryomyces hansenii* and *Pichia anomala* facilitate the alcoholic fermentation (Table 2.2). Although the local method of *pito* production shares similarities with other traditional alcoholic beverages, its fermentation is unique in that there are distinct changes in pH during fermentation, with an initial increase from 4.2 to 5.2 within 24 hours, followed by a sharp decrease to 3.7 after 48 hours (Kolawole et al., 2007).

2.3 Fermented Cereal-Based Products from East Africa

2.3.1 Injera

Injera, a traditional Ethiopian bread, is a leavened, flat round bread typically made from cereals such as teff (*Eragrostis tef* (Zucc) Trotter) and sorghum (Anyango et al., 2011), or various cereal combinations, including teff and white sorghum, wheat and red sorghum, and barley and wheat (Baye et al., 2013). Making *injera* involves mixing teff flour (or a combination of flours) with water and allowing it to ferment for several days, similar to a sourdough starter (Figure 2.2). This fermentation imparts a mildly sour taste to the *injera*. Anberbir et al. (2023) documented a process for *injera* preparation. It involves starter (*ersho*) preparation, batter preparation, fermentation (primary

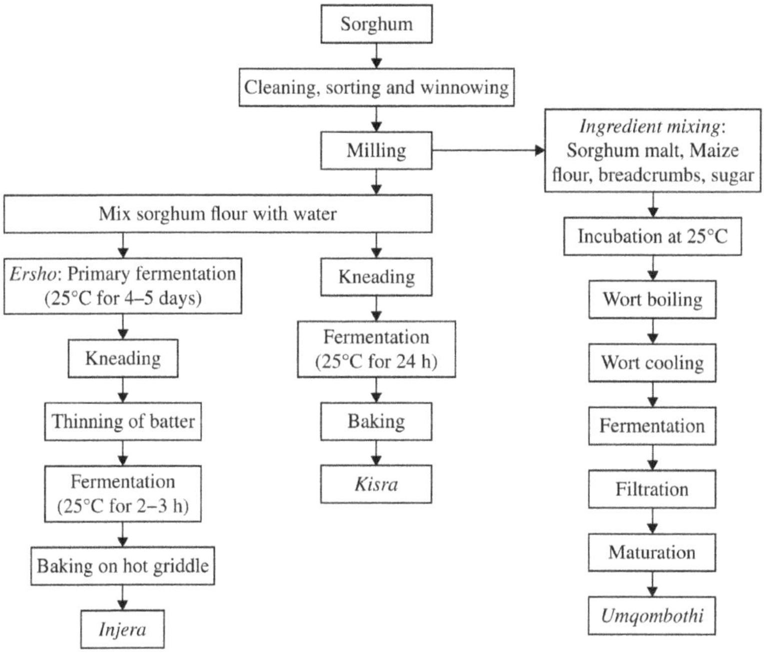

Figure 2.2 Flow chart of the production of some sorghum-based African fermented food products from Africa.

and secondary) and preparation of *injera* from batter by baking. The *ersho* is an undefined starter culture used for Ethiopian *injera* and is prepared by combining water and teff flour in a 2:1 ratio and leaving the mixture to ferment for up to 72 hours without any mixing. The *injera* batter is made by combining the teff flour and water in a ratio of 1:2 and mixing this with 160 mL of *ersho*. Then, the mixture is kneaded for 5 minutes by hand, covered with a lid and kept aside at room temperature (25°C), with fermentation allowed for 36–108 hours as indicated in the experimental design. This fermentation step is considered to be the primary fermentation. The microorganisms in the *ersho* are mainly yeasts (namely, *Saccharomyces cerevisiae* and *Kazachstania humilis*) and LAB strains (*Enterococcus lactis*, *Lactocaseibacillus paracasei* and *Lactobacillus* species) (Dessalegn & Andualem, 2023a, 2023b).

After this step, 10% of the fermented mass (from the primary fermentation) is taken for the preparation of *absit*, which serves as a binder and is prepared with a ratio of 1:3 v/v (batter: water) by boiling on a hot plate and then cooling to a temperature of 45°C (Yegrem et al., 2021). It is then added back to the remaining fermented batter from the primary fermentation, mixed carefully and allowed 2–3 hours for secondary fermentation. The resulting batter is used for the *injera* preparation.

Once fermented, the *injera* batter is baked into large, flat pancakes (Anberbir et al., 2023). This can be done using a specialised electric stove or, more commonly, on a clay plate known locally as Amharic *mittad* or Tigrinya *mogogo*, which is placed over a fire. In terms of its shape, *injera* resembles the French crêpe and the South Indian dosa, as it is a circular-shaped flatbread used as a base for other foods. The bottom surface of the *injera*, which comes into contact with the heating surface, typically has a smooth texture, while the top surface becomes porous. This porous structure allows consumers to use the *injera* to scoop up sauces and dishes effectively (Abiyu et al., 2013).

2.3.2 Busaa

Busaa is a type of opaque maize beer originating from Kenya, resembling *malwa* beer in Uganda and *kaffir* beer in South Africa (Aka et al., 2014). Across various African countries, numerous variations of opaque beers exist, each with its local name. Opaque maize beer is often considered nutritionally superior to clear lager beer due to its higher levels of crude protein, thiamine and riboflavin. The preparation of *busaa* typically uses maize endosperm grits and finger-millet malt (*Eleusine coracana*) with a process similar to that of making kaffir beer (Farnworth, 2003).

Upon completion, *busaa* typically contains around 0.5–1% lactic acid and alcohol (2–4% v/v) (Gélinas & McKinnon, 2000; Nout, 1980). During the maize souring stage, the fermentation is driven by LAB and a few yeasts, predominantly *Lactobacillus helveticus*, *Lactobacillus salivarius*, *Pediococcus damnosus*, *Pediococcus parvulus*, *Candida krusei* and *Saccharomyces cerevisiae*. The primary fermentation also involves a combination of yeasts and LAB, including *Candida krusei* and *Lacticaseibacillus casei* var. *rhamnosus*. While the production techniques of *kaffir* beer and *busaa* share similarities, there are some key differences. In *busaa*, corn souring precedes the addition of malt, whereas in *kaffir* beer, souring occurs at approximately 50°C, creating favourable conditions for *Lactobacilli* and excluding yeasts (Gélinas & McKinnon, 2000). In *kaffir* beer, the wort resulting from maize saccharification by sorghum malt is boiled for 2 hours before inoculation with previously manufactured beer, a step that eliminates LAB from the souring stage. *Busaa* fermentation is spontaneous and uncontrolled, resembling the production process of *pito*.

2.4 Fermented Cereal-Based Products from Northern Africa

2.4.1 *Hussuwa*

Hussuwa holds significance as one of the key traditional fermented foods in Sudan. Its classification varies across different country regions, where it may be considered a solid food or a beverage (El Sheikha & Montet, 2014). Traditionally, the production of *hussuwa* relies on natural fermentation by indigenous microorganisms. It is made from a semi-solid mixture of sorghum flour and sorghum malt in a ratio of 1:0.5. This mixture is left to ferment for approximately 12 hours, after which it is lightly cooked as dough or as pancakes. During the fermentation of sorghum grains to produce *hussuwa*, several key microorganisms involved in the fermentation process, including LAB (*Lm. fermentum*, *P. acidilactici*, *Lb. saccharolyticum*) and yeast (*Saccharomyces cerevisiae*), play essential roles (El Sheikha & Montet, 2014; Franz & Holzapfel, 2019). Throughout this process, both lactic and ethanolic fermentation occur, resulting in a final product with a sweet–sour taste (El Sheikha & Montet, 2014).

2.4.2 *Kisra*

Kisra is a flatbread made from sorghum or millet flour, fermented with naturally occurring LAB and yeast. It is typically baked into thin, round sheets, measuring approximately 1–1.5 mm thick and 30–45 cm in diameter. Ideally, *kisra* should have a supple, soft and moist texture without being spongy (Ali & Mustafa, 2009). There is considerable potential for *kisra* to develop into a gluten-free sandwich wrap (AwadElkareem & Taylor, 2011). The traditional fermentation process of *kisra* involves mixing sorghum or millet flour with water in a ratio of about 1:2 (w/v). Generally, a starter from a previous fermentation batch, known as the mother dough, is added at a level of approximately 10%. This technique is called back-slopping. Fermentation typically takes around 12–19 hours, during which time the pH decreases from approximately 6.0 to less than 4.0 (Ali & Mustafa, 2009). The fermented dough is then baked into thin sheets and served with various types of stew made from vegetables and meat (El Sheikha & Montet, 2014). *Lm. fermentum*, *Lb. reuteri*, *Lb. confusus*, *Lb. amylovorus*, *Lv. brevis* and *Pediococcus pentosaceus* have been identified as the predominant microorganisms involved in *kisra* fermentation (Ali & Mustafa, 2009; Eltayeb et al., 2020). Additionally, other microorganisms such as *Enterococcus faecium* and *Saccharomyces cerevisiae* have also been identified in *kisra* (Abdel Rahman et al., 2010).

2.4.3 Balady

Balady bread is a traditional staple food in Egypt, also known as Egyptian flatbread. It is characterised by its round, flat shape and unique texture, which is both soft and chewy (Mohsen et al., 2016). Its production involves a relatively simple yet intricate process of several key steps such as mixing, fermenting, shaping, proofing and baking. The primary ingredients for *balady* bread are whole wheat flour, water, salt and a leavening agent which is typically a natural sourdough or commercial yeast (Abd-El-Khalek et al., 2014; Azouz et al., 2010; Mohsen et al., 2016).

Fermentation is a crucial stage in the preparation of *balady* bread. The dough is left to ferment for 2–3 hours, allowing *Saccharomyces cerevisiae* to metabolise sugars and produce carbon dioxide. This process not only helps in leavening the dough but also enhances the flavour and texture of the bread. The traditional use of sourdough starter introduces beneficial LAB such as *Lactiplantibacillus plantarum* into the dough, which contributes to the bread's unique tangy taste and extended shelf life, due to the production of organic acids (Mohsen et al., 2016). After fermentation, the dough is divided into portions and shaped into round, flat discs. These discs are then left to proof, or rise, for a short period (around 30 minutes). During proofing, the dough continues to ferment, further developing its structure and flavour. The dough is typically baked in a very hot oven, often at temperatures exceeding 250°C. Traditional baking methods include the use of a taboon or a brick oven, which imparts a distinctive smoky flavour and texture to the bread. The high baking temperature causes the bread to puff up quickly, creating a pocket in the middle, characteristic of this type of flatbread.

2.5 Fermented Cereal-Based Products from Southern Africa

2.5.1 Mahewu

Mahewu, also known as *amahewu*, *mageu*, is a sour beverage made from maize meal and consumed in various African countries and some Arabian Gulf regions (Chavan et al., 1989). This traditional beverage serves as adult food and typical weaning food for children. In South Africa it is referred to by different names. Zulus call it *amahewu*, Xhosas call it *amarehwu*, Swazis call it *emahewu*, Pedis call it *metogo*, Sothos call it *machleu* and Vendas call it *maphulo*. However, the most commonly used term is *mahewu*.

Mahewu is typically prepared by mixing maize porridge with water and adding sorghum, millet malt or wheat flour (Pswarayi & Gänzle, 2019, 2022). The mixture is then left to ferment spontaneously at ambient temperature. Alternatively, leftover pap can be mashed into a slurry and fermented overnight. The microbiology of *mahewu* fermentation involves the spontaneous action of microorganisms naturally present in the environment. LAB are predominant during fermentation, particularly species such as *Lactococcus lactis* subsp. *lactis* (Idowu et al., 2016; Pswarayi and Gänzle, 2019). These LAB contribute to the acidification of the beverage, resulting in its characteristic sour taste.

Traditional brewing methods are often employed, particularly in township communities, where *mahewu* is brewed using age-old techniques. While commercial production may involve controlled fermentation processes, natural fermentation is common in traditional settings. Fermentation is essential for the flavour development, preservation and nutritional enhancement of *mahewu*. Understanding the microbial dynamics of *mahewu* fermentation is crucial for maintaining traditional brewing practices and ensuring product quality and safety.

2.5.2 Umqombothi

Umqombothi, a traditional fermented beer widely popular among the Black South African community, is characterised by its pink, opaque appearance, yoghurt-like flavour, thin consistency and slight effervescence, with an alcohol content of around 3% (Katongole, 2008). Typically consumed in its actively fermenting state, umqombothi has a short shelf life of two to three days (Aka et al., 2014; Katongole, 2008). Despite advancements in production techniques, traditional brewing methods persist in township communities, where women brew *umqombothi* using age-old techniques. Maize and sorghum, often combined, are the primary cereals employed in *umqombothi* production (Figure 2.2).

The key microorganisms in *umqombothi* fermentation include LAB and yeasts. The LAB, such as *Apilactobacillus*, *Ligilactobacillus*, *Lactiplantibacillus*, *Lactobacillus* and *Lactococcus* species, play a crucial role in acidification, contributing to the sour flavour and aiding in preservation. Yeasts, including species like *Saccharomyces cerevisiae*, are responsible for alcohol production, leading to the beverage's low alcohol content (Hlangwani et al., 2020, 2023). The fermentation process occurs at ambient temperature, allowing the indigenous microflora in the raw materials to initiate fermentation spontaneously. While commercial production may involve controlled fermentation processes, traditional methods often rely on natural fermentation, which contributes to the unique flavour profile of *umqombothi*. The microbiology of *umqombothi* is integral to its fermentation process, influencing its flavour, texture and shelf life. Understanding the microbial dynamics involved in *umqombothi* fermentation is essential for preserving traditional brewing practices and ensuring product quality and safety.

Fermented cereal-based foods are a crucial part of the indigenous knowledge base of foods on the African continent, with diverse products peculiar to each region. A pictorial view of selected traditional cereal-based fermented food products from various regions of Africa is depicted in Figure 2.3.

Figure 2.3 A pictorial view of selected traditional cereal-based fermented food products from different regions of Africa and the distinct fermenting microorganisms. (From *Pito* (Zaukuu et al., 2016), *kisra* (https://shorturl.at/Qj3M1), *balady* (https://shorturl.at/X2zNS), *injera* (https://shorturl.at/qppgA), *bussaa* (https://shorturl.at/jKSDP), *umqombothi* (https://shorturl.at/GmDSR), *mahewu* (https://rb.gy/zzg7be) and *kenkey* (https://tinyurl.com/8n7m33w9).)

2.6 Conclusion

The microbiology of African fermented cereals is a complex and diverse field that encompasses various microorganisms and fermentation techniques. Traditional fermentation processes involve the spontaneous fermentation of cereals such as maize, sorghum and millet, producing staple foods like *ogi*, *kenkey*, *injera*, *umqombothi* and *kisra*. Key microorganisms involved in these fermentation processes include LAB, yeasts and moulds, which contribute to flavour development, preservation and the nutritional enhancement of the final products. Future trends in the microbiology of African fermented cereals will likely focus on the characterisation of microbial communities using advancements in molecular techniques, such as next-generation sequencing, which will enable a more comprehensive understanding of the microbial communities involved in African fermented cereals. This will provide insights into the diversity, dynamics and functional roles of microorganisms during fermentation. There is a need for more research to explore the functional and health-promoting properties of fermented cereals, including their probiotic potential, bioactive compound production and impact on gut microbiota composition. This will inform the development of functional foods with enhanced nutritional and health benefits.

There is a growing interest in optimising traditional fermentation processes to improve product quality, safety and consistency. Additionally, there is a trend towards developing novel fermentation techniques, such as controlled fermentation, starter culture inoculation and bioreactor-based fermentation, to meet the demands of modern food processing and consumer preferences. Future research will focus on using African fermented cereals as ingredients in novel food products and formulations, including baked goods, snacks and beverages. This will create opportunities for value addition, waste reduction and the development of sustainable food systems. Overall, the microbiology of African fermented cereals holds great promise for the production of nutritious, culturally important and commercially viable food products. Continued research efforts and technological advancements will drive innovation and foster the sustainable development of traditional fermented foods in Africa and beyond.

References

Abd-El-Khalek, M. H., Abd El-Kader, M. H., & Elkholany, E. A. (2014). Modeling of the changes in freshness, microflora and sensory attributes of Egyptian balady bread stored under different temperatures. *Middle East Journal of Applied Sciences*, *4*(3), 674–683.

Abdel Rahman, I. A., Hamad, S. H., Osman, M. A., & Dirar, H. A. (2010). Characterization and distribution of microorganisms associated with kisra bread preparation from three sorghum varieties in Sudan. *Current Research in Bacteriology*, *3*, 138–147.

Abiyu, H. T., Woldegiorgis, A. Z., & Haki, G. (2013). Preparation of injera from pre-fermented flour: Nutritional and sensory quality. *International Journal of Science Innovations and Discoveries*, *3*(1), 165–175.

Abriouel, H., Omar, N. B., López, R. L., Martínez-Cañamero, M., Keleke, S., & Gálvez, A. (2006). Culture-independent analysis of the microbial composition of the African traditional fermented foods poto poto and dégué by using three different DNA extraction methods. *International Journal of Food Microbiology*, *111*(3), 228–233.

Adebo, O. A. (2020). African sorghum-based fermented foods: Past, current and future prospects. *Nutrients*, *12*(4), 1111.

Adebo, O. A., Chinma, C. E., Obadina, A. O., Soares, A. G., Panda, S. K., & Gan, R.-Y. (2023). An insight into indigenous fermented foods for the tropics. In Adebo, O. A., Chinma, C. E., Obadina, O., Panda, S., Soares, A., & Gan, R.Y. (Eds.), *Indigenous fermented foods in the tropics* (pp. 1–12). Elsevier.

Adegbehingbe, K. T., Adeleke, B. S., & Adejoro, D. (2019). Microbiological assessment, physico-chemical and functional agidi produced in Akoko area of Ondo state. *FJPAS*, *2*(1), 275282.

Afolayan, A. O., Ayeni, F. A., & Ruppitsch, W. (2017). Antagonistic and quantitative assessment of indigenous lactic acid bacteria in different varieties of ogi against gastrointestinal pathogens. *The Pan African Medical Journal*, *27*.

Aka, S., Konan, G., Fokou, G., Dje, K. M., & Bonfoh, B. (2014). Review on African traditional cereal beverages. *American Journal of Research Communication*, *2*(5), 103–153.

Ali, A. A., & Mustafa, M. M. (2009). Use of starter cultures of lactic acid bacteria and yeasts in the preparation of kisra, a Sudanese fermented food. *Pakistan Journal of Nutrition*, *8*(9), 1349–1353.

Ampe, F., & Miambi, E. (2000). Cluster analysis, richness and biodiversity indexes derived from denaturing gradient gel electrophoresis fingerprints of bacterial communities demonstrate that traditional maize fermentations are driven by the transformation process. *International Journal of Food Microbiology*, *60*(1), 91–97.

Amusa, N. A., & Ashaye, O. A. (2009). Effect of processing on nutritional, microbiological and sensory properties of kunun-zaki (a sorghum based non-alcoholic beverage) widely consumed in Nigeria. *Pakistan Journal of Nutrition*, *8*(3), 288–292.

Anberbir, S. M., Satheesh, N., Abera, A. A., Tenagashaw, M. W., Asres, D. T., Tiruneh, A. T., Habtu, T. A., Awol, S. J., & Wudineh, T. A. (2023). Effect of blending ratio and fermentation time on the physicochemical, microbiological, and sensory qualities of injera from teff, pearl millet, and buckwheat flours. *CyTA – Journal of Food*, *21*(1), 217–236.

Annan, T. (2013). *Development of starter culture for the fermentation of dehulled maize into* nsiho *(white* kenkey*)* [Doctoral dissertation, University of Ghana, Legon].

Anumudu, C. K., Omeje, F. I., & Obinwa, G. N. (2018). Microbial succession pattern in Ogi fermentation. *International Journal of Advanced Research in Biological Sciences*, *5*(7), 247–251.

AwadElkareem, A. M., & Taylor, J. R. (2011). Protein quality and physical characteristics of kisra (fermented sorghum pancake-like flatbread) made from tannin and non-tannin sorghum cultivars. *Cereal Chemistry*, *88*(4), 344–348.

Anyango, J. O., de Kock, H. L., & Taylor, J. R. (2011). Impact of cowpea addition on the Protein Digestibility Corrected Amino Acid Score and other protein quality parameters of traditional African foods made from non-tannin and tannin sorghum. *Food Chemistry*, *124*(3), 775–780.

Azouz, A., Hafez, S. A., & Soliman, S. A. (2010). Evaluation of balady bread made with sour dough compared to that made by bakery yeast. *Egyptian Journal of Agricultural Research*, *88*(2), 543–556.

Baye, K., Mouquet-Rivier, C., Icard-Vernière, C., Rochette, I., & Guyot, J. P. (2013). Influence of flour blend composition on fermentation kinetics and phytate hydrolysis of sourdough used to make injera. *Food Chemistry*, *138*(1), 430–436.

Chavan, J. K., Kadam, S. S., & Beuchat, L. R. (1989). Nutritional improvement of cereals by fermentation. *Critical Reviews in Food Science & Nutrition*, *28*(5), 349–400.

Chelule, P. K., Mokoena, M. P., & Gqaleni, N. (2010). Advantages of traditional lactic acid bacteria fermentation of food in Africa. *Current Research, Technology and Education Topics in Applied Microbiology and Microbial Biotechnology*, *2*, 1160–1167.

Dessalegn, M., & Andualem, B. (2023a). Isolation and screening of yeast starters from ersho samples collected from different parts of Ethiopia and determination of their potential role for bread production. *International Journal of Food Properties*, *26*(2), 3030–3047.

Dessalegn, M., & Andualem, B. (2023b). In vitro evaluation of probiotic properties of yeasts and lactic acid bacteria isolated from ersho samples collected from different parts of Ethiopia. *International Journal of Food Properties*, *26*(2), 3048–3064.

Diaz, M., Kellingray, L., Akinyemi, N., Adefiranye, O. O., Olaonipekun, A. B., Bayili, G. R., Ibezim, J., Du Plessis, A. S., Houngbe´dji, M., Kamya, D., & Mukisa, I. M. (2019). Comparison of the microbial composition of African fermented foods using amplicon sequencing. *Scientific Reports*, *9*(1), 18.

Djameh, C., Ellis, W. O., Oduro, I., Saalia, F. K., Blay, Y. M., & Komlaga, G. A. (2019). West African sorghum beer fermented with *Lactobacillus delbrueckii* and *Saccharomyces cerevisiae*: Shelf-Life and consumer acceptance. *Journal of the Institute of Brewing*, *125*(3), 333–341.

Djameh, C., Saalia, F. K., Sinayobye, E., Budu, A., Essilfie, G., Mensah-Brown, H., & Sefa-Dedeh, S. (2015). Optimization of the sorghum malting process for pito production in Ghana. *Journal of the Institute of Brewing*, *121*(1), 106–112.

Efiuvwevwere, B. J. O., & Akona, O. (1995). The microbiology of 'Kunun-zaki', a cereal beverage from Northern Nigeria, during the fermentation (production) process. *World Journal of Microbiology and Biotechnology*, *11*(5), 491–493.

Ekpa, O., Palacios-Rojas, N., Kruseman, G., Fogliano, V., & Linnemann, A. R. (2019). Sub-Saharan African maize-based foods-processing practices, challenges and opportunities. *Food Reviews International*, *35*(7), 609–639.

El Sheikha, A. F., & Montet, D. (2014). African fermented foods: Historical roots and real benefits. In *Microorganisms and fermentation of traditional foods* (pp. 248–282). CRC Press.

Eltayeb, M. M., Eltigani, S. A., & Taniguchi, T. (2020). Pyrosequencing scrutiny of bacterial and fungal communities in two Sudanese sorghum-based fermented foods. *Annals of Microbiology*, *70*, 1–10.

Fadahunsi, I. F., & Soremekun, O. O. (2017). Production, nutritional and microbiological evaluation of mahewu a South African traditional fermented porridge. *Journal of Advances in Biology & Biotechnology*, *14*, 1–10.

Farnworth, E. R. (2003). *Handbook of fermented functional foods*. CRC Press.

Franz, C. M., & Holzapfel, W. H. (2019). Examples of lactic-fermented foods of the African continent. In *Lactic acid bacteria* (pp. 235–253). CRC Press.

Gélinas, P., & McKinnon, C. (2000). Fermentation and microbiological processes in cereal foods. In *Handbook of cereal science and technology, revised and expanded* (pp. 741–754). CRC Press.

Greppi, A., Rantsiou, K., Padonou, W., Hounhouigan, J., Jespersen, L., Jakobsen, M., & Cocolin, L. (2013). Determination of yeast diversity in ogi, mawè, gowè and tchoukoutou by using culture-dependent and-independent methods. *International Journal of Food Microbiology*, *165*(2), 84–88.

Halm, M., Amoa-Awua, W. K., & Jakobsen, M. (2004). Kenkey: An African fermented maize product. In *Handbook of food and beverage fermentation technology* (pp. 940–959). CRC Press.

Hlangwani, E., Abrahams, A., Masenya, K., & Adebo, O. A. (2023). Analysis of the bacterial and fungal populations in South African sorghum beer (umqombothi) using full-length 16S rRNA amplicon sequencing. *World Journal of Microbiology and Biotechnology*, *39*(12), 350.

Hlangwani, E., Adebiyi, J. A., Doorsamy, W., & Adebo, O. A. (2020). Processing, characteristics and composition of umqombothi (a South African traditional beer). *Processes*, *8*(11), 1451.

Houngbédji, M., Johansen, P., Padonou, S. W., Akissoé, N., Arneborg, N., Nielsen, D. S., Hounhouigan, D. J, & Jespersen, L. (2018). Occurrence of lactic acid bacteria and yeasts at species and strain level during spontaneous fermentation of mawè, a cereal dough produced in West Africa. *Food Microbiology*, *76*, 267–278.
Idowu, O. O., Fadahunsi, I. F., & Onabiyi, O. A. (2016). Production and nutritional evaluation of *Mahewu*: A non-alcoholic fermented beverage of South Africa. *International Journal of Research in Pharmacy and Biosciences*, ***3***(6), 27–33.
Ignat, M. V., Salanță, L. C., Pop, O. L., Pop, C. R., Tofană, M., Mudura, E., Coldea, T. E., Borșa, A., & Pasqualone, A. (2020). Current functionality and potential improvements of non-alcoholic fermented cereal beverages. *Foods*, *9*(8), 1031.
Ire, F. S., Edio, P. A., & Maduka, N. (2020). Comparative assessment of the microbiological and physicochemical quality of a laboratory brewed 'burukutu' and commercialized products sold in some markets in Port Harcourt Nigeria. *European Journal of Biology and Biotechnology*, *1*(5), 114.
Katongole, J. N. (2008). *The microbial succession in indigenous fermented maize products* [Doctoral dissertation, University of the Free State].
Kayitesi, E., Onojakpor, O., & Moyo, S. M. (2023). Highlighting the impact of lactic-acid-bacteria-derived flavours or aromas on sensory perception of African fermented cereals. *Fermentation*, *9*(2), 111.
Kolawole, O. M., Kayode, R. M. O., & Akinduyo, B. (2007). Proximate and microbial analyses of burukutu and pito produced in Ilorin, Nigeria. *African Journal of Biotechnology*, *6*(5), 587.
Konfo, C. T. R., Chabi, N. W., Dahouenon-Ahoussi, E., Cakpo-Chichi, M., Soumanou, M. M., & Sohounhloue, D. C. K. (2015). Improvement of African traditional sorghum beers quality and potential applications of plants extracts for their stabilization: A review. *Journal of Microbiology, Biotechnology and Food Sciences*, *5*, 190196.
Lei, V., & Jakobsen, M. (2004). Microbiological characterization and probiotic potential of koko and koko sour water, African spontaneously fermented millet porridge and drink. *Journal of Applied Microbiology*, *96*(2), 384–397.
Lyumugabe, F., Kamaliza, G., Bajyana, E., & Thonart, P. H. (2010). Microbiological and physico-chemical characteristic of Rwandese traditional beer "Ikigage". *African Journal of Biotechnology*, *9*(27), 4241–4246.
Mashau, M. E., Maliwichi, L. L., & Jideani, A. I. O. (2021). Non-alcoholic fermentation of maize (*Zea mays*) in sub-Saharan Africa. *Fermentation*, *7*(3), 158.
Merabti, R., Bekhouche, F., Chuat, V., Madec, M. N., Maillard, M. B., Bailly, S., Thierry, A., & Valence, F. (2015). A large diversity of lactic acid bacteria species is involved in the fermentation of wheat used for the manufacture of *lemzeiet*. *European Food Research and Technology*, *241*, 137–149.
Mohsen, S. M., Aly, M. H., Attia, A. A., & Osman, D. B. (2016). Effect of sourdough on shelf life, freshness and sensory characteristics of Egyptian balady bread. *Journal of Applied & Environmental Microbiology*, *4*(2), 39–45.

Mukisa, I. M., Porcellato, D., Byaruhanga, Y. B., Muyanja, C. M., Rudi, K., Langsrud, T., & Narvhus, J. A. (2012). The dominant microbial community associated with fermentation of Obushera (sorghum and millet beverages) determined by culture-dependent and culture-independent methods. *International Journal of Food Microbiology*, *160*(1), 1–10.

Neela, S., & Fanta, S. W. (2020). Injera (An ethnic, traditional staple food of Ethiopia): A review on traditional practice to scientific developments. *Journal of Ethnic Foods*, *7*(1), 115.

Nout, M. J. R. (1980). Process development and preservation of Busaa, a Kenyan traditional opaque maize beer. *Chemie, Mikrobiologie, Technologie der Lebensmittel*, *6*, 175–182.

Nwosu, V. C., & Oyeka, C. A. (1998). Microbiological succession occurring during fermentation of ogi—An African breakfast cereal. *Journal of the Elisha Mitchell Scientific Society*, *114*(4), 190–198.

Nyanzi, R., & Jooste, P. J. (2012). Cereal-based functional foods. *Probiotics*, *8*, 161–197.

Oguntoyinbo, F. A., & Narbad, A. (2012). Molecular characterization of lactic acid bacteria and in situ amylase expression during traditional fermentation of cereal foods. *Food Microbiology*, *31*(2), 254–262.

Oguntoyinbo, F. A., Tourlomousis, P., Gasson, M. J., & Narbad, A. (2011). Analysis of bacterial communities of traditional fermented West African cereal foods using culture independent methods. *International Journal of Food Microbiology*, *145*(1), 205–210.

Omemu, A. M., Oyewole, O. B., & Bankole, M. O. (2007). Significance of yeasts in the fermentation of maize for ogi production. *Food Microbiology*, *24*(6), 571–576.

Onyango, C., Okoth, M. W., & Mbugua, S. K. (2000). Effect of drying lactic fermented uji (an East African sour porridge) on some carboxylic acids. *Journal of the Science of Food and Agriculture*, *80*(13), 1854–1858.

Oyarekua, M. A., & Eleyinmi, A. F. (2004). Comparative evaluation of the nutritional quality of corn, sorghum and millet ogi prepared by a modified traditional technique. *Journal of Food Agriculture and Environment*, *2*, 94–99.

Oyedeji, O., Ogunbanwo, S. T., & Onilude, A. A. (2013). Predominant lactic acid bacteria involved in the traditional fermentation of fufu and ogi, two Nigerian fermented food products. *Food and Nutrition Sciences*, *4*(11), 40.

Pswarayi, F., & Gänzle, M. (2022). African cereal fermentations: A review on fermentation processes and microbial composition of non-alcoholic fermented cereal foods and beverages. *International Journal of Food Microbiology*, *378*, 109815.

Pswarayi, F., & Gänzle, M. G. (2019). Composition and origin of the fermentation microbiota of mahewu, a Zimbabwean fermented cereal beverage. *Applied and Environmental Microbiology*, *85*(11), e03130–18.

Sanata, B., Adama, Z., Ibrahim, S., Apoline, S., Mamoudou, C., Constant, S., Robert, G. T., & Christophe, H. (2017). Characterization of the fungal flora of dolo, a traditional fermented beverage of Burkina Faso, using MALDI-TOF mass spectrometry. *World Journal of Microbiology and Biotechnology*, *33*(9), 1–5.

Sawadogo-Lingani, H., Lei, V., Diawara, B., Nielsen, D. S., Møller, P. L., Traoré, A. S., & Jakobsen, M. (2007). The biodiversity of predominant lactic acid bacteria in *dolo* and *pito* wort, for production of sorghum beer. *Journal of Applied Microbiology*, *103*, 765–777.

Soro-Yao, A. A., Brou, K., Amani, G., Thonart, P., & Djè, K. M. (2014). The use of lactic acid bacteria starter cultures during the processing of fermented cereal-based foods in West Africa: A review. *Tropical Life Sciences Research*, *25*(2), 81.

Tadesse, B. T., Abera, A. B., Tefera, A. T., Muleta, D., Alemu, Z. T., & Wessel, G. (2019). Molecular characterization of fermenting yeast species from fermented Teff dough during preparation of injera using ITS DNA sequence. *International Journal of Food Science*, *2019*.

Tano, M. B., Aka-Gbezo, S., Attchelouwa, C. K., & Koussémon, M. (2020). Use of lactic acid bacteria as starter cultures in the production of Tchapalo, a traditional sorghum beer from Côte d'Ivoire. *American Journal of Food Science and Health*, *6*, 23–31.

Tekle, B., Jabasingh, S. A., Fantaw, D., Gebreslassie, T., Rao, S. R. M., Baraki, H., & Gebregziabher, K. (2019). An insight into the Ethiopian traditional alcoholic beverage: Tella processing, fermentation kinetics, microbial profiling and nutrient analysis. *LWT – Food Science and Technology*, *107*, 9–15.

Tokpohozin, S. E., Waldenmaier, J. F. T., Fischer, S. E., & Becker, T. (2017). Polyphasic characteristics of lactic acid bacteria isolated from Beninese sorghum beer starter. *LWT – Food Science and Technology*, ***80***, 51–58.

Turpin, W., Humblot, C., Noordine, M. L., Wrzosek, L., Tomas, J., Mayeur, C., Cherbuy, C., Guyot, J. P., & Thomas, M. (2013). Behavior of lactobacilli isolated from fermented slurry (ben-saalga) in gnotobiotic rats. *PLoS ONE*, *8*(4), e57711.

Vieira-Dalodé, G., Jespersen, L., Hounhouigan, J., Moller, P. L., Nago, C. M., & Jakobsen, M. (2007). Lactic acid bacteria and yeasts associated with gowe production from sorghum in Benin. *Journal of Applied Microbiology*, *103*(2), 342–349.

Yegrem, L., Abera, S., & Temesgen, M. (2021). Nutritional composition and sensory quality of injera prepared from teff (*Eragrostis teff* (Zucc.) Trotter) complemented with lupine (*Lupinus* spp.). *Cogent Food & Agriculture*, *7*(1), 1862469.

Zaukuu, J. L. Z., Oduro, I., & Ellis, W. O. (2016). Processing methods and microbial assessment of pito (an African indigenous beer), at selected production sites in Ghana. *Journal of the Institute of Brewing*, *122*(4), 736–744.

3

Microbiology of African Fermented Legumes, Pulses and Oil Seeds

3.1 Introduction

The fermentation of various indigenous African legumes, pulses and oilseeds is known as alkaline fermentation (Odunfa & Oyewole, 1998; Ugwuanyi & Okpara, 2019). Alkaline fermentation primarily pertains to the fermentation of legumes (such as soybeans) and protein-rich oilseeds (e.g., African locust bean and African oil beans) to yield condiments. Through these processes, the pH typically rises to 8 or higher. This pH elevation is ascribed to the metabolic actions of microorganisms that degrade the proteins in the raw materials into peptides, amino acids and ammonia (Falade & Akinrinde, 2021; Ouoba et al., 2003). This solid-state fermentation technique for African legumes, pulses and oil seeds is commonly employed in various African countries (particularly in West and Central Africa) to produce condiments or seasoning agents that play a significant role in flavour enhancement and the nutritional enrichment of African dishes. Table 3.1 depicts key African fermented foods derived from legumes, pulses and oilseeds.

In traditional African food processing, fermentation is a widely used technique that relies on the activity of various microorganisms to transform raw ingredients into stable and flavourful foods. The microbiology of alkaline fermentation involves a diverse range of microorganisms which contribute to the fermentation process. One of the key microorganisms involved is the *Bacillus* species, including *Bacillus subtilis* and *Bacillus cereus* (Gernah et al., 2007; Oguntoyinbo et al., 2007). These bacteria can thrive in alkaline environments and play a crucial role in the breakdown of complex carbohydrates, proteins and lipids present in the raw ingredients. Through their enzymatic activities, *Bacillus*

DOI: 10.1201/9781003658320-3

Table 3.1 Key African Fermented Foods and Beverages from Legumes, Pulses and Oilseeds

PRODUCT	RAW MATERIAL USED	FOOD USES	COUNTRY/REGION
Afitin	African locust bean	Condiment	Benin
Datou	Roselle seed	Condiment	Mali
Dawadawa	African locust bean/ soybean/Bambara groundnut	Condiment	Central and West Africa
Furundu	Roselle seed	Condiment	Sudan
Iru	African locust bean	Condiment	West Africa
Kinda	African locust bean	Condiment	Sierra Leone
Nététou	African locust bean	Condiment	Senegal
Ogiri/Ogiri-igbo	Melon	Condiment	Central, East and West Africa
Ogiri-nwan	Castor oil seed	Condiment	Nigeria
Ogiri-ugu	Fluted pumpkin bean	Condiment	Nigeria
Ogboroti	African oil bean and melon seed	Condiment and delicacy	Nigeria
Okpehe/Okpiye	Mesquite	Condiment	Nigeria
Owoh	Cottonseed	Condiment	Nigeria
Sigda/Ogiri-saro	Sesame seed	Condiment	Sudan and Sierra Leone
Soumbala	African locust bean	Condiment	Burkina Faso
Ugba	African oil bean	Salad	Nigeria
Tunjanee	Groundnut seed	Meat analogue	Sudan

Source: Adapted from Adebo et al. (2023).

species contribute to the degradation of anti-nutritional factors such as phytates and tannins, thereby improving the nutritional quality of the fermented products (Oguntoyinbo et al., 2007).

In addition to *Bacillus* species, other bacteria such as lactic acid bacteria (LAB) may also be present in alkaline-fermented foods, particularly in soybean-based products (Musliu et al., 2021; Omafuvbe et al., 2002). The LAB contributes during the onset of the fermentation process by producing organic acids, such as lactic acid, which help lower the fermentation medium's pH and inhibit the growth of spoilage organisms. Furthermore, LAB plays a role in flavour development and contributes to the fermented products' characteristic flavour and aroma. Occasionally, yeasts and moulds may also be present in alkaline fermented foods; however, their presence and contribution to the fermentation process may vary depending on the specific type of fermented food and the fermentation conditions. Most of these condiments are products in which *Bacillus* species are the key organisms. Table 3.2 summarises the

Table 3.2 Major Fermentation Microbiota of Common African Fermented Products from Oil Legumes, Pulses and Oilseeds

PRODUCT	RAW MATERIAL	PRODUCT FORM	FERMENTATION MICROORGANISMS	COUNTRY	REFERENCES
Dawadawa, Iru, Afitin, Sonru, Soumbala, Nététou and Kinda	African locust bean seed	Condiment	*Lysinibacillus irui* sp. nov., *Bacillus pumilus, Bacillus megaterium, Bacillus licheniformis, Bacillus subtilis, Bacillus cereus, Bacillus* spp., *Leuconostoc mesenteroides, Leuconostoc dextranicum*	Nigeria, Ghana, Benin, Burkina Faso, Senegal, Sierra Leone	Azokpota et al. (2006), Ouoba et al. (2008), Agbobatinkpo et al. (2013), Adamu et al. (2018) and Akintayo et al. (2023)
Soy-Dawadawa,	Soybean	Condiment	*B. licheniformis, B. pumilus, B. subtilis*	Nigeria, Burkina Faso, Ghana	Titilayo and Musliu (2016) and Omafuvbe et al. (2000)
Dawadawa type	Bambara groundnut	Condiment	*B. licheniformis, B. pumilus, B. subtilis* subsp. *subtilis, Bacillus amyloliquefaciens* subsp. *plantarum*		Akanni et al. (2018a) and Akanni et al. (2018b)
Bikalga, yayanku/ikpiru, mbuja, dawadawa botso, datou, furundu	Roselle seeds		*B. cereus, B. licheniformis, B. megaterium, B. pumilus, Bacillus badius, Bacillus sphaericus, Bacillus fusiformis*	Niger, Mali, Sudan	Ouoba et al. (2008)
Ugba	African oil bean	Salad/condiment	*Bacillus subtilis, B. licheniformis, B. sphaericus, B. pumilus, Bacillus firmus, Bacillus* spp., *Pediococcus* spp., *Enterococcus* spp.	Nigeria	Olasupo et al. (2016) and Ogbulie et al. (2014)

(Continued)

Table 3.2 Major Fermentation Microbiota of Common African Fermented Products from Oil Legumes, Pulses and Oilseeds *(Continued)*

PRODUCT	RAW MATERIAL	PRODUCT FORM	FERMENTATION MICROORGANISMS	COUNTRY	REFERENCES
Ogiri	Melon seed, castor oil seed, fluted pumpkin seed	Condiment	*B. pumilus*, *B. licheniformis*, *B. subtilis*, *Lactobacillus* species, *Enterococcus faecalis*	Nigeria	Egbebi et al. (2016) and Barber et al. (1989)
Owoh	Cottonseed	Condiment	*B. licheniformis*, *B. pumilus*, *Bacillus* species	Nigeria	Sanni and Ogbonna (1991) and Olasupo and Okorie (2019)
Okpehe/Okpiye	*Prosopis africana* seed	Condiment	*Bacillus* spp., *Lactobacillus* spp., *Pseudomonas* spp.	Nigeria	Achi (2005)
Sigda	Sesame seedcake	Meat-like substitute	*Streptococcus* spp., *Pediococcus* spp., *Candida* spp., *Saccharomyces* spp., *Debaromyces* spp.	Western Sudan	Elfaki et al. (1991) and Harper and Collins (1992)
Tunjanee/Omtigania	Groundnut seedcake	Meat-like substitute	*Bacillus* spp., *Propionibacterium*, *Lactobacillus* spp.	Sudan	ElGasim et al. (2012)

microbiota of some common alkaline-fermented African condiments that are key players in traditional food systems across parts of Africa. Understanding the role of these microorganisms is essential for optimising fermentation processes, ensuring product safety and quality and harnessing the nutritional benefits of traditional African fermented foods.

3.2 Selected African Fermented Legume, Pulses and Oil Seed-Based Products

3.2.1 Dawadawa

Dawadawa, a traditional African condiment predominantly found in West and Central Africa, is identified by several names such as *iru*, *soumbala*, *nététou*, *afitin*, *sonru* and *kinda* (Agbobatinkpo et al., 2019). *Dawadawa* is a condiment derived through fermentation of the African locust bean (*Parkia biglobosa*), while variants of *dawadawa*-type condiments are also produced from other oilseeds, such as the soybean (*Glycine max*) and Bambara groundnut (*Vigna subterranean*) (Akanni et al., 2018a; Omafuvbe et al., 2000). For *afitin* and *sonru* production, an additional starter culture carrier is included to aid fermentation. *Dawadawa* holds great significance in African cuisines, serving as a food flavouring or seasoning agent in soups, stews, sauces and marinades. Its popularity is due to the volatile compounds and amino acids produced during fermentation (Akanni et al., 2018b; Amoa-Awua et al., 2014; Onyenekwe et al., 2014). In addition to its flavour-enhancing properties, *dawadawa* plays a substantial role in providing essential nutrients such as protein, essential fatty acids and B-vitamins, with a particular emphasis on riboflavin (Achi, 2005). The processing and fermentation steps of dawadawa are summarised in Figure 3.1.

The fermentation typically occurs in an alkaline environment, with the pH rising to 8 or above due to microbial activities breaking down proteins into peptides, amino acids and ammonia. This transformation yields a flavourful and aromatic condiment rich in umami taste compounds. *Bacillus* species, particularly the *B. subtilis* group, tend to be predominant during the fermentation of legumes and oilseeds to produce *dawadawa* (Akanni et al., 2018a; Ezeokoli et al., 2016; Jeyaram et al., 2011). *B. subtilis*, *Bacillus pumilus*, *Bacillus megaterium*, *Bacillus amyloliquefaciens*, *Bacillus licheniformis*, *Bacillus sphaericus*, *Bacillus fusiformis*, *Bacillus badius* and *B. cereus* are frequently identified in the fermentation

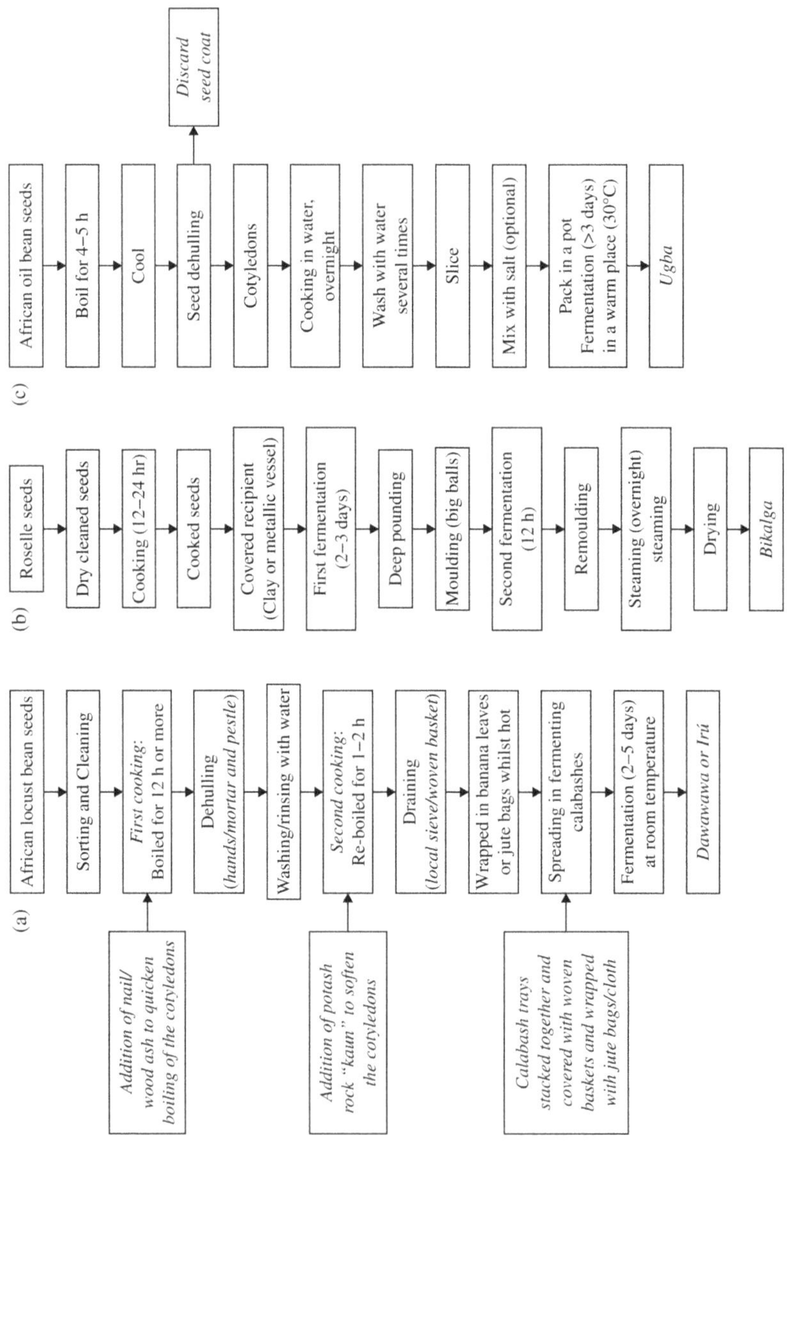

Figure 3.1 Flow chart of the production of (a) *dawadawa*, (b) *bikalga* and (c) *ugba*.

process. *Bacillus* species counts reaching 10 Log10 CFU/g in the final products have been reported (Kaboré et al., 2012; Owusu-Kwarteng et al., 2022; Parkouda et al., 2009; Parkouda et al., 2010). A novel species of *Lysinibacillus irui* sp. nov. isolated from *iru* has been reported (Akintayo et al., 2023). In *soy-dawadawa*, *Bacillus* species identified included *B. amyloliquefaciens*, *B. licheniformis*, *Bacillus mojavensis*, *B. pumilus*, *B. subtilis*, *B. subtilis* subsp. *spizizenii* and *Bacillus tequilensis* (Ezeokoli et al., 2016, 2018; Omafuvbe et al., 2002).

In a *dawadawa*-type condiment made from Bambara groundnut, *B. subtilis* subsp. *subtilis*, *B. amyloliquefaciens* subsp. *plantarum*, *B. pumil* and *B. licheniformis* were the predominant species (Akanni et al., 2018a, 2018b). These *Bacillus* species play a crucial role in breaking down proteins into smaller peptides and amino acids through enzymatic activity (Odunfa, 1985; Ouoba et al., 2003), thereby contributing to the development of flavour compounds and enhancing the nutritional quality of the final product. Additionally, the free amino acids are utilised by the *Bacillus* as carbon and nitrogen sources, to produce ammonia/ammonium hydroxide, resulting in high pH values and the associated atypical odour of these fermented condiments (Amoa-Awua et al., 2014; Ouoba et al., 2005). Degradation of anti-nutritive factors and improvements in the digestibility of the fermented seed condiments have also been reported (Ugwuanyi & Okpara, 2019).

3.2.2 Bikalga

Bikalga (Burkina Faso), also known as *mbuja* (Cameroon), *dawadawa-botso* or *Mari-Bi* (Niger), *fururndu* (Sudan), *datou* (Mali) and *yayanku/ikpiru* (Benin), is made from the fermentation of roselle (*Hibiscus sabdariffa*) (Compaoré et al., 2013; Rabiou et al., 2019; Tahir et al., 2021). Processing and fermentation of roselle seeds differ depending on traditional knowledge and the intended use of the final product. For instance, the Beninese *yanyanku* and *ikpiru* are not directly consumed as condiments; they are mainly used as functional additives and play a vital role as enrichment inocula or starter culture carriers to African locust bean fermentation for the production of *dawadawa* (Agbobatinkpo et al., 2011, 2012, 2013; Azokpota et al., 2011). Typically, the fermentation of *bikalga* begins with soaking the dehulled roselle seeds in water to soften them, then cooking the soaked seeds, followed by grinding or pounding to form a paste. This paste is then left to ferment in ambient conditions for several days to weeks, depending on the desired characteristics of the final product (Figure 3.1). During fermentation, a diverse microbial community develops, predominantly *Bacillus* species such as *B. subtilis*, *B. licheniformis* and *B. amyloliquefaciens*, in an alkaline fermentation brought about by the extensive hydrolysis of the proteins into amino acids and volatiles (Agbobatinkpo et al., 2019; Compaoré et al., 2020). Analysis of the variants *yanyanku* and *ikpiru* has found that the *Bacillus* genera predominates, mainly *B. subtilis*, *B. cereus*, *B. amyloliquefaciens*, *B. licheniformis*, *Bacillus safensis* and *Bacillus altitudinis*. Other species that occur sporadically during fermentation are *Bacillus aryabhattai*, *Bacillus flexus*, *Bacillus circulans* (0–2%) and species of the genera *Lysinibacillus*, *Paenibacillus*, *Brevibacillus* and *Aneurinibacillus* (Agbobatinkpo et al., 2013). The final pH values of both additives range between 6.2 and 10, and *Bacillus* spore counts vary between 5.5 and 8.9 Log10 CFU/g (Agbobatinkpo et al., 2011). These microorganisms are responsible for the fermentation process, during which they metabolise carbohydrates, proteins and lipids present in the raw material. *Bacillus* species are known for their proteolytic and lipolytic activities, which contribute to the breakdown of proteins and lipids into simpler compounds such as peptides, amino acids and fatty acids. While the fermentation process primarily involves beneficial microorganisms responsible for transforming raw materials into the desired product, there is a possibility of pathogenic microbes being present, albeit in small numbers. These pathogenic microbes could potentially include bacteria such as *Salmonella* spp., *Escherichia coli*, *Staphylococcus aureus* and *Clostridium* spp., as well as moulds and yeasts that may produce harmful toxins (Agbobatinkpo et al., 2019). The complex interaction between these microorganisms leads to unique sensory attributes and a preserved, fermented condiment.

3.2.3 Ugba

Ugba is a traditional condiment produced through the fermentation of sliced African oil bean (*Pentaclethra macrophylla* Benth) seeds. It boasts up to 44% protein content, encompassing at least 17 of the 20 essential amino acids (Enujiugha et al., 2003; Mbah et al., 2018). Production of *ugba* is through alkaline fermentation, with a pH ranging from

5.65 to 8.20 (Ogueke et al., 2015). Through fermentation, microorganisms initiate biochemical reactions, breaking down complex carbohydrates, proteins and lipids into simpler compounds, resulting in the characteristic flavour, texture and aroma of *ugba* (Anosike et al., 2022). The digestibility and utilisation of the protein of the seeds are also improved (Anosike et al., 2022; Okechukwu et al., 2015). To prepare *ugba*, oil bean seeds are initially boiled for 3 hours, dehulled and cooked. The cooked seeds are sliced thinly (0.5–1 mm thick), boiled again for 2 hours, drained, rinsed three times and then soaked in cold water for 4 hours to remove the bitter taste. Subsequently, the sliced beans are wrapped in banana leaves, packed in a clean container and left to ferment at room temperature (Figure 3.1).

The fermentation process of *ugba* comprises natural fermentation in banana leaves or other suitable containers. During fermentation, a consortium of microorganisms proliferates on the seeds, including bacteria and fungi which are present in the environment. The fermentation process usually takes several days to weeks, depending on environmental conditions and desired product characteristics. It has been reported that *Bacillus* species are the most predominant microorganisms during fermentation. Other organisms that play a minimal role include *Staphylococcus* spp., *Micrococcus* sp., *Leuconostoc* species and fungi (Ogbulie et al., 2014; Ogueke et al., 2015).

The *Bacillus* species count will continue to increase until the end of the fermentation process. At the same time, other organisms are found to decrease in number after 24 hours (Ogueke et al., 2015). *Bacillus* species identified in *ugba* include *B. licheniformis*, *B. sphaericus*, *B. subtilis*, *B. pumilus* and *Bacillus firmus* (Ahaotu et al., 2013; Ogbulie et al., 2014). These microorganisms play crucial roles during fermentation by breaking down proteins and lipids in the raw material. *Bacillus* species are known for their ability to produce proteolytic and lipolytic enzymes, which hydrolyse proteins and lipids into simpler compounds such as peptides, amino acids and fatty acids. The other microorganisms present in *ugba* are due to chance inoculation from the environment. LAB may be present, contributing to the fermentation process by producing organic acids and contributing to the flavour development of the final product. Identified fungi species include *Mucor* spp., *Penicillium* spp. and *Aspergillus* spp. (Okechukwu et al., 2015) which could cause spoilage. The synergistic action of these microorganisms results in the desirable sensory attributes and the preservation of *Ugba* during fermentation. Once fermentation is complete, the *Ugba* is ready for consumption or further processing into various dishes.

3.2.4 *Sigda* and *Tunjanee/Omtigania*

Sigda is a meaty-tasting traditional Sudanese condiment made from sesame (*Sesamum indicum L.*) oilseed cake. The fermentation involves the action of LAB and yeasts that transform the substrate into a flavourful and nutritious product (Elfaki et al., 1991; Harper & Collins, 1992). Typically, the fermentation begins with the preparation of the sesame seedcake. The cake is then mixed with water and pounded into a paste (Figure 3.1). The paste is packed into an appropriate container and left to ferment for 168 hours (7 days). After fermentation, the fermented *sidga* are moulded into small balls and dried (Elfaki et al., 1991). The microorganisms during fermentation tend to be dominated by homofermentative *Streptococcus* spp. and *Pediococcus* spp. Although initially the yeasts *Candida* spp. and *Saccharomyces* spp. are also present, the LAB species dominate the fermentation (Elfaki et al., 1991; Harper & Collins, 1992). Bacteria such as LAB are commonly involved in the fermentation process, contributing to the production of organic acids, enzymes and flavour compounds. Yeasts may also play a role in fermentation by producing ethanol and other secondary metabolites. Throughout the fermentation period, which can vary in duration, the microbial community interacts and metabolises the substrate, leading to the development of the unique flavours, textures and nutritional properties characteristic of *Sigda*. It is worth noting that the *sigda* fermentation is quite different from that of traditional oilseed fermentations practised in West Africa with respect to both the microflora and the products of fermentation (Elfaki et al., 1991).

Tunjanee or *Omtigania* is a Sudanese meat substitute made from groundnut (*Arachis hypogea*) seedcake. The fermentation process is similar to *Sigda*, in that the fermentation of the groundnut cake takes 168 hours, and it is dominated by *Streptococcus* spp. and *Lactobacillus* spp. (Ali & Elkhalil, 2022; ElGasim et al., 2012). During the initial fermentation stages, consortia of microflora consisting of LAB, yeasts, mould and acetic acid bacteria are present (Ali & Elkhalil, 2022). The production and accumulation of acids decrease the pH of the fermenting mixture from 5.7 to 3.6 at the end of fermentation.

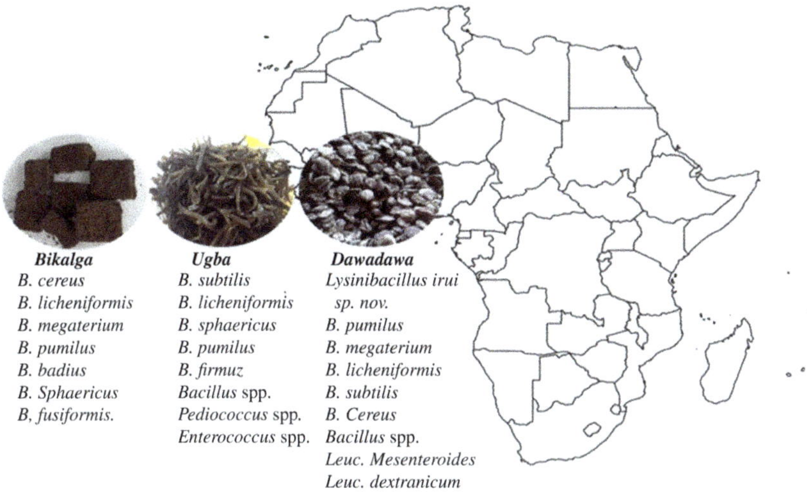

Figure 3.2 A pictorial view of selected traditional African fermented legume-based food products and the distinct fermenting microorganisms. (*Bikalga/furundu* (From Al Bahi & Abdelgadir, 2012), *dawadawa* (From Chinma et al., 2023) and *ugba* (https://tinyurl.com/yew7fzkk).)

The *Streptococcus* spp. and *Lactobacillus* spp. are the microbial populations that dominate the entire fermentation process (Ali & Elkhalil, 2022). Selected traditional legume-based fermented food products from various regions of Africa are depicted in Figure 3.2.

3.3 Conclusion

The traditional fermentation of various indigenous African legumes, pulses and oilseeds is an intriguing process driven by the metabolic activities of diverse microorganisms. Alkaline fermentation, which typically involves *Bacillus* species and sometimes LAB, plays a pivotal role in transforming raw ingredients into the stable and flavourful condiments that are essential components of African cuisine. Throughout this fermentation process, complex carbohydrates, proteins and lipids are broken down into simpler compounds, particularly peptides, amino acids and fatty acids, contributing to the development of unique sensory characteristics and enhancing the nutritional quality of the final products. *Bacillus* species, particularly *B. subtilis* and related strains, are predominant in many alkaline-fermented African condiments such as *dawadawa*, *ugba*, *bikalga*, *sigda* and *tunjanee*. These bacteria exhibit proteolytic and lipolytic activities, which are crucial for the breakdown of proteins and lipids, leading to the formation of flavour compounds and improving product digestibility. Additionally, LAB may contribute to the fermentation process by producing organic acids, lowering the pH and enhancing the safety and sensory attributes of the fermented foods. Overall, understanding the microbiology of alkaline fermentation is essential for optimising fermentation processes, ensuring product safety and quality and harnessing the nutritional benefits of traditional African fermented foods. Further research into the microbial dynamics and biochemical pathways comprising these fermentation processes could provide valuable insights into developing novel food products and preserving cultural culinary traditions.

References

Achi, O. K. (2005). Traditional fermented protein condiments in Nigeria. *African Journal of Biotechnology*, *4*(13).

Adamu, S. A., Farouq, A. A., Magashi, M. A., & Sokoto, A. M. (2018). Microorganisms associated with the traditional fermentation of African locust beans for the production of tasty food condiment. *African Journal of Agriculture and Food Science*, *1*, 12–18.

Adebo, O. A., Chinma, C. E., Obadina, A. O., Soares, A. G., Panda, S. K., & Gan, R.-Y. (2023). An insight into indigenous fermented foods for the tropics. In Adebo, O. A., Chinma, C. E., Obadina, O., Panda, S., Soares, A., & Gan, R.Y. (Eds.), *Indigenous fermented foods in the tropics* (pp. 1–12). Elsevier.

Agbobatinkpo, B. P., Azokpota, P., Akissoe, N., Kayod, P., da Gbadji, R., & Hounhouigan, D. J. (2011). Indigenous perception and characterization of Yanyanku and Ikpiru: Two functional additives for the fermentation of African locust bean. *Ecology of Food and Nutrition*, *50*(2), 101–114.

Agbobatinkpo, B. P., Tossou, G. M., Adinsi, L., Akissoe, H. N., & Hounhouigan, D. J. (2019). Optimal fermentation parameters for processing high quality African locust bean condiments. *Journal of Food Science and Technology*, *56*, 4648–4657.

Agbobatinkpo, P. B., Dabade, S. D., Laleye, F., Akissoe, N., Azokpota, P., & Hounhouigan, J. D. (2012). Softening effect of Ikpiru and Yanyanku, two traditional additives used for the fermentation of African Locust Bean (*Parkia biglobosa*) seeds in Benin. *International Journal of Biological and Chemical Sciences*, *6*(3), 1281–1292.

Agbobatinkpo, P. B., Thorsen, L., Nielsen, D., Azokpota, P., Akissoe, N., Hounhouigan, D. J., & Jakobsen, M. (2013). Biodiversity of aerobic endospore-forming bacterial species occurring in Yanyanku and Ikpiru, fermented seeds of *Hibiscus sabdariffa* used to produce food condiments in Benin. *International Journal of Food Microbiology*, *163*, 231–238.

Ahaotu, I., Anyogu, A., & Njoku, H. O. (2013). Molecular identification and safety of *Bacillus* species involved in the fermentation of African oil beans (*Pentaclethra macrophylla* Benth) for production of *ugba*. *International Journal of Food Microbiology*, *162*(1), 95–104.

Akanni, G. B., De Kock, H. L., Naudé, Y., & Buys, E. M. (2018b). Volatile compounds produced by *Bacillus* species alkaline fermentation of bambara groundnut (*Vigna subterranean* (L.) Verdc) into a dawadawa-type African food condiment using headspace solid-phase microextraction and GC× GC–TOFMS. *International Journal of Food Properties*, *21*(1), 930–942.

Akanni, G. B., Naudé, Y., de Kock, H. L., & Buys, E. M. (2018a). Diversity and functionality of *Bacillus* species associated with alkaline fermentation of bambara groundnut (*Vigna subterranean* L. Verdc) into dawadawa-type African condiment. *European Food Research and Technology*, *244*, 1147–1158.

Akintayo, S. O., Neumann, B., Fischer, M., Henkel, M., Lilge, L., & Hausmann, R. (2023). *Lysinibacillus irui* sp. nov., isolated from Iru, fermented African locust beans. *International Journal of Systematic and Evolutionary Microbiology*, *73*(11), 006167.

Al Bahi, A. A., & Abdelgadir, W. S. (2012). Effects of dietary furundu (fermented *Hibiscus sabdariffa* seed) supplementation on the performance, some blood and serum parameters and histopathology of Wistar rats. *Advance Journal of Food Science and Technology*, *4*(2), 64–69.

Ali, S. E. M., & Elkhalil, E. A. I. (2022). A note on microbiological and biochemical evaluation of 'Omtigania', a Sudanese fermented food based on peanut seed cake. *University of Khartoum Journal of Agricultural Sciences*, *30*(1), 104–104.

Amoa-Awua, W. K., Awusi, B., Owusu, M., Appiah, V., Ofori, H., Thorsen, L., & Jespersen, L. (2014). Reducing the atypical odour of dawadawa: Effect of modification of fermentation conditions and post-fermentation treatment on the development of the atypical odour of dawadawa. *Food Control*, *42*, 335–342.

Anosike, F. C., Onyemah, K. O., Ossai, C. U., Ofoegbu, J. G., Okpaga, F. O., Ikpeama, C. C., Nkwegu, F. M., Nwankwo, S. C., Onyeji, G. N., Inyang, P., & Ndifon, E. M. (2022). Probiotic potential and viability of bacteria in fermented African oil bean seed (*Pentaclethra macrophylla*): A mini review. *Applied Food Research*, *2*(1), 100082.

Azokpota, P., Houndenoukon, M. S., Hounhouigan, J. D., Nago, M. C., & Jakobsen, M. (2011). Evaluation of yanyanku processing, an additive used as starter culture to produce fermented food condiments in Benin. *Journal of Food Processing and Preservation*, *35*(6), 813–821.

Azokpota, P., Hounhouigan, D. J., & Nago, M. C. (2006). Microbiological and chemical changes during the fermentation of African locust bean *Parkia biglobosa*, to produce Afitin, Iru and Sonru, three traditional condiments produced in Benin. *International Journal of Food Microbiology*, *107*, 304–309.

Barber, L. A., Ibrahim, E. M., & Achinewhu, S. C. (1989). Microorganisms associated with fluted pumpkin (*Telfairia occidentalis*). *International Journal of Science and Technology*, *24*, 189–194.

Chinma, C. E., Ezeocha, V. C., Adedeji, O. E., Inyang, C. U., Enujiugha, V. N., & Adebo, O. A. (2023). African legume, pulse, and oilseed-based fermented products. In *Indigenous fermented foods for the tropics* (pp. 73–84). Academic Press.

Compaoré, C. S., Nielsen, D. S., Sawadogo-Lingani, H., Berner, T. S., Nielsen, K. F., Adimpong, D. B., Diawara, B., Ouédraogo, G. A., Jakobsen, M., & Thorsen, L. (2013). Bacillus amyloliquefaciens ssp. plantarum strains as potential protective starter cultures for the production of Bikalga, an alkaline fermented food. *Journal of applied microbiology*, *115*(1), 133–146.

Compaoré, C. S., Parkouda, C., Traoré, E. M., Savadogo, A., & Jespersen, L. (2020). Development of starter cultures carrier for the production of high quality soumbala, a food condiment based on *Parkia biglobosa* seeds. *African Journal of Biotechnology*, *19*(11), 820–828.

Egbebi, A. O., Seidu, K. T., & Muhammed, A. A. (2016). Nutritional and microbiological analyses of fermented locust bean (*Parkia biglobosa*) and fermented melon (*Citrullus vulgaris*). *Savant Journal of Agricultural Research*, *2*, 1–6.

Elfaki, A. E., Dirar, H. A., Collins, M. A., & Harper, D. B. (1991). Biochemical and microbiological investigations of sigda—A Sudanese fermented food derived from sesame oilseed cake. *Journal of the Science of Food and Agriculture*, *57*(3), 351–365.

ElGasim, A., Yagoub, A. E., & Ahmed, A. T. (2012). Physicochemical and Microbiological study on tunjanee—A traditionally fermented Sudanese food from groundnut (*Arachis hypogaea*) seed cake. *Global Advanced Research Journal of Food Science and Technology*, *1*, 008–017.

Enujiugha, V. N., Badejo, A. A., Iyiola, S. O., & Oluwamukomi, M. O. (2003). Effect of germination on the nutritional and functional properties of African oil bean (*Pentaclethra macrophylla* Benth) seed flour. *Journal of Food Agriculture and Environment*, *1*, 72–75.

Ezeokoli, O., Gupta, A., Popoola, T., & Bezuidenhout, C. (2016). Molecular analysis of bacterial community dynamics during the fermentation of soy-daddawa condiment. *Food Science and Biotechnology*, *25*, 1081–1086.

Ezeokoli, O. T., Adeleke, R. A., & Bezuidenhout, C. C. (2018). Core bacterial community of soy-daddawa: Insights from high-throughput DNA metabarcoding. *Lwt*, *97*, 61–66.

Falade, K. O., & Akinrinde, I. M. (2021). Physical, chemical and adsorption isotherm characteristics of fermented soybean cultivars, and cracked and dehulled African locust bean using selected *Bacillus* spp. *Journal of Food Science and Technology*, *58*, 2749–2760.

Gernah, D. I., Inyang, C. U., & Ezeora, N. L. (2007). Incubation and fermentation of African locust beans (*Parkia biglobosa*) in production of "dawadawa". *Journal of Food Processing and Preservation*, *31*(2), 227–239.

Harper, D. B., & Collins, M. A. (1992). Leaf and seed fermentations of Western Sudan. In Gaden, E. L., Bokanga, M., Harlander, S., Hesseltine, C. W., Steinkraus, K. H. (eds), *Applications of Biotechnology in Traditional Fermented Foods* (pp. 105–113). National Academic Press. Washington DC.

Jeyaram, K., Romi, W., Singh, T. A., Adewumi, G. A., Basanti, K., & Oguntoyinbo, F. A. (2011). Distinct differentiation of closely related species of *Bacillus subtilis* group with industrial importance. *Journal of Microbiological Methods*, *87*(2), 161–164.

Kaboré, D., Thorsen, L., Nielsen, D. S., Berner, T. S., Sawadogo-Lingani, H., Diawara, B., Dicko, M. H., & Jakobsen, M. (2012). Bacteriocin formation by dominant aerobic sporeformers isolated from traditional maari. *International Journal of Food Microbiology*, *154*(1–2), 10–18.

Mbah, G. O., Onyeabo, U. A., & Udeh, B. C. (2018). Effect of fermentation on nutritional composition of African oil bean seed. *The Pacific Journal of Science and Technology*, *19*(1), 244–250.

Musliu, A., Yusuf, M., & Adebisi, S. (2021). In vivo evaluation of soya beans flour fermented with lactic acid bacteria as a potential probiotic food. *Frontiers in Microbiology*, *31*(4), 16–27.

Odunfa, S. A. (1985). Biochemical changes in fermenting African locust bean (Parkia biglobosa) during 'iru' fermentation. *International Journal of Food Science & Technology*, *20*(3), 295–303.

Odunfa, S. A., & Oyewole, O. B. (1998). African fermented foods. In B. J.W. Wood (Ed.), *Microbiology of Fermented Foods*, (2nd ed., Vol. 2, pp. 713–752). Blackie Academic and Professional.

Ogbulie, T. E., Nsofor, C. A., & Nze, F. C. (2014). Bacteria species associated with ugba (*Pentaclethra macrophylla* Bentham) produced traditionally and in the laboratory and the effect of fermentation on the product of oligosaccharide hydrolysis. *Nigerian Food Journal*, *32*(2), 73–80.

Ogueke, C. C., Anosike, F., & Owuamanam, C. I. (2015). Prediction of amino nitrogen during ugba (*Pentaclethra macrophylla*) production under different fermentation variables. A response surface approach. *Nigerian Food Journal*, *33*, 61–66.

Oguntoyinbo, F. A., Sanni, A. I., Franz, C. M. A. P., & Holzapfel, W. H. (2007). Phenotypic diversity and technological properties of *Bacillus subtilis* species isolated from okpehe, a traditional fermented condiment. *World Journal of Microbiology and Biotechnology*, *23*, 401–410.

Okechukwu, E. C., Hah, A. Y., & Aguora, S. O. (2015). Effect of varying fermentation temperature on the microbial population of African oil bean seed (*Pentaclethra macrophylla* Benth). *Journal of Environmental Science, Toxicology and Food Technology*, *9*(6), 34–37.

Olasupo, N. A., Okorie, C. P., & Oguntoyinbo, F. A. (2016). The biotechnology of ugba, a Nigerian traditional fermented food condiment. *Frontiers in Microbiology*, *7*, 1153, Article.

Olasupo, N. A., & Okorie, P. C. (2019). African fermented food condiments: Microbiology impacts on their nutritional values. In R. L. Solis-Oviedo, & A. de la Cruz Pech-Canul (Eds.), *Frontiers and new trends in the science of fermented food and beverage*. IntechOpen.

Omafuvbe, B. O., Abiose, S. H., & Shonukan, O. O. (2002). Fermentation of soybean (*Glycine max*) for soy-daddawa production by starter cultures of *Bacillus*. *Journal of Food Microbiology*, *19*, 561–566.

Omafuvbe, B. O., Shonukan, O. O., & Abiose, S. H. (2000). Microbiological and biochemical changes in the traditional fermentation of soybean for 'soy-daddawa'—Nigerian food condiment. *Food Microbiology*, *17*(5), 469–474.

Onyenekwe, P. C., Odeh, C., & Enemali, M. O. (2014). Identification and quantification of headspace volatile constituents of okpehe, fermented *Prosopis africana* seeds. *International Food Research Journal*, *21*(3), 1193–1197.

Ouoba, L. I. I., Diawara, B., Annan, N. T., Poll, L., & Jakobsen, M. (2005). Volatile compounds of Soumbala, a fermented African locust bean (*Parkia biglobosa*) food condiment. *Journal of Applied Microbiology*, *99*(6), 1413–1421.

Ouoba, L. I. I., Parkouda, C., Diawara, B., Scotti, C., & Varnam, A. H. (2008). Identification of *Bacillus* spp. from Bikalga, fermented seeds of *Hibiscus sabdariffa*: phenotypic and genotypic characterization. *Journal of applied microbiology*, *104*(1), 122–131.

Ouoba, L. I. I., Rechinger, K. B., Barkholt, V., Diawara, B., Traore, A. S., Jakobsen, M., et al. (2003). Degradation of proteins during the fermentation of African locust bean (*Parkia biglobosa*) by strains of *Bacillus subtilis* and *Bacillus pumilus* for Soumbala. *Journal of Applied Microbiology*, *94*, 396–402.

Owusu-Kwarteng, J., Parkouda, C., Adewumi, G. A., Ouoba, L. I. I., & Jespersen, L. (2022). Technologically relevant *Bacillus* species and microbial safety of West African traditional alkaline fermented seed condiments. *Critical Reviews in Food Science and Nutrition*, *62*(4), 871–888.

Parkouda, C., Nielsen, D. S., Azokpota, P., Ivette Irène Ouoba, L., Amoa-Awua, W. K., Thorsen, L., … & Jakobsen, M. (2009). The microbiology of alkaline-fermentation of indigenous seeds used as food condiments in Africa and Asia. *Critical Reviews in Microbiology*, *35*(2), 139–156.

Parkouda, C., Thorsen, L., Compaoré, C. S., Nielsen, D. S., Tano-Debrah, K., Jensen, J. S., … & Jakobsen, M. (2010). Microorganisms associated with Maari, a Baobab seed fermented product. *International Journal of Food Microbiology*, *142*(3), 292–301.

Rabiou, M. M., Lewamy, M., Tchicama, M. M., Alma, M. M., & Sadou, H. (2019). Manufacturing processes and physicochemical analyzes of Mari-bi, a Niger Republic traditional spice obtained by *Hibiscus sabdariffa* seeds fermentation. *Journal de la Société Ouest-Africaine de Chimie*, *48*, 43–48.

Sanni, A. I., & Ogbonna, D. N. (1991). The production of owoh—A Nigerian fermented seasoning agent from cotton seed (*Gossypium hirsitium* L.). *Food Microbiology*, *8*, 223–229.

Tahir, H. E., Mahunu, G. K., Xiaobo, Z., & Mariod, A. A. (2021). Ethnobotanical uses, fermentation studies and indigenous preferences of *Hibiscus sabdariffa*. In: *Roselle (Hibiscus sabdariffa)* (pp. 231–254). Academic Press.

Titilayo, A., & Musliu, A. (2016). Biochemical changes and sensory evaluation of soy iru produced using starter culture. *British Microbiology Research Journal*, *14*(6), 1–10.

Ugwuanyi, J. O., & Okpara, A. N. (2019). Current status of alkaline fermented foods and seasoning agents of Africa. In *New advances on fermentation processes*. Intech Open. https://doi.org/10.5772/intechopen.87052

4
Microbiology of African Fermented Dairy

4.1 Introduction

Fermented dairy products have been an integral part of African culinary traditions for centuries, reflecting the continent's diverse cultures, climates and culinary practices (Agyei et al., 2020; Benkerroum, 2013). These traditional foods play significant roles in nutrition, preservation and cultural rituals, contributing to the social fabric of African communities. The fermentation of dairy products in Africa encompasses a wide range of techniques, ingredients and microbial communities, resulting in a rich tapestry of flavours, textures and nutritional properties (Agyei et al., 2020).

In many African societies, dairy products hold cultural and symbolic importance and are often incorporated into religious ceremonies, rites of passage and social gatherings. Fermented dairy products are prized for their role in traditional medicine, being believed to promote digestive health and overall well-being. These products are deeply embedded in culinary traditions, with each region boasting its own unique recipes and preparation methods passed down through generations (Bell et al., 2023). Africa's vast geographical and cultural diversity is reflected in the wide array of fermented dairy products found across the continent. Each region, from West Africa to East Africa, and North Africa to Southern Africa, has distinct fermented dairy specialities. In Southern Africa, traditional fermented milk products such as *amasi* are widely consumed among communities such as the Zulu and Xhosa, serving as a source of nutrition and a symbol of heritage (Maleke et al., 2021). Similarly, in Northern Africa, fermented dairy products such as *laban* or *rayeb* are popular, providing essential nutrients and flavours unique to the region (Samet-Bali et al.,

DOI: 10.1201/9781003658320-4

2016). In East Africa, communities such as the Maasai are known for consuming fermented camel's milk, often referred to as *Kule-naoto* or *suusac*, which plays a central role in rituals and ceremonies (Jans et al., 2012; Mathara et al., 2004). In West Africa, fermented milk products like *nunu* or *nono* are common among communities such as the Fulani, contributing to dietary diversity and cultural identity (Akabanda et al., 2013; Walsh et al., 2017). These fermented dairy products are valued for their nutritional content, probiotic properties and a longer shelf life than fresh milk. These products may include yoghurt-type products, cheeses, butters and beverages, each with its own unique flavour profile and texture. A variety of African fermented dairy products is summarised in Table 4.1. Commonly used dairy sources include cow, goat, sheep, camel and buffalo milk, depending on local availability and cultural preferences (Mattiello et al., 2018; Owusu-Kwarteng et al., 2020).

Table 4.1 African Dairy-Based Fermented Products

PRODUCT	MILK SOURCE	PRODUCT FORM	COUNTRY/REGION
Aguat	Cow	Fermented whey	Ethiopia
Amabere	Cow	Gruel	Kenya
Amasi/Mafi	Cow	Gruel	Southern Africa
Arera	Cow	Buttermilk	Ethiopia
Ayib	Cow	Cottage cheese	Ethiopia
Bouhezza	Goat	Raw goat cheese	Algerian
Chambiko	Cow	Beverage	Malawi
Chekapmkaika/ Kwerionik	Cow	Beverage	Uganda
Cuku	Cow	Cheese	Nigeria
Dhanaan	Camel	Beverage	Ethiopia
Ergo	Cow, goat and sheep	Beverage	Ethiopia
Fene	Cow	Beverage	Mali
Gariss	Camel	Beverage	Sudan, Somalia
Gibna Bayda/Gibna Madaffra	Cow	Cheese	Sudan
Ititu	Cow	Curd	Ethiopia
Jben	Cow	Cheese	Morocco
Kibe	Cow	Butter	Ethiopia
Kindirmo	Cow	Sour yoghurt	Cameroon, Nigeria
Kiviguto/Makamo	Cow	Beverage, dessert	Rwanda, Uganda

(Continued)

Table 4.1 African Dairy-Based Fermented Products *(Continued)*

PRODUCT	MILK SOURCE	PRODUCT FORM	COUNTRY/REGION
Klila	Cow	Cheese	Algeria
Kule-naoto	Cow	Beverage	Kenya
Kwerionik	Cow	Beverage	Uganda
Laban	Cow	Gruel	Egypt
Lait caille	Cow	Beverage	Burkina Faso, Côte d'Ivoire, Senegal
Leben/Lben	Cow	Sour milk	East Central and North Africa
Mabisi	Cow	Gruel	Zambia
Madila	Goat	Beverage, curd	Botswana
Maishanu	Cow	Butter	Nigeria
Masse	Cow	Homemade yoghurt	Mozambique
Mursik	Cow or goat	Beverage	Kenya
Nitir kibe	Cow	Ghee	Ethiopia
Nunu	Cow	Beverage	Ghana, Nigeria
Nyarmie	Cow	Curd	Ghana
Omarere	Cow	Beverage	Namibia
Omashikawa	Cow	Buttermilk	Namibia
Pendidaam	Cow	Beverage	Cameroon
Rayeb	Cow or camel	Curdled skim milk	Egypt
Roub/Rob	Camel	Beverage	Sudan
Sithemi	Cow	Beverage	South Africa
Smen	Cow	Butter	Morocco
Suusac	Camel	Beverage	Kenya, Somalia
Zabady	Buffalo, cow, or a mixture of both	Beverage	Egypt
Zebda beldi	Camel	Butter	North Africa

Source: Adapted from Mourad and Bettache (2019) and Agyei et al. (2020).

In general, the production of fermented milk products relies on the age-old traditional method of natural fermentation, widely practised across Africa. This involves allowing raw or heated milk to ferment naturally, or through a technique known as 'back-slopping', where a portion of a successfully fermented previous batch is used as a starter to inoculate a new batch (Agyei et al., 2020; Owusu-Kwarteng et al., 2017). The microbiota present in African dairy-based fermented products is diverse, incorporating the rich nutrients present in milk and milk products. Milk comprises sugars such as lactose, glucose and galactose, which can be metabolised by microorganisms to produce organic acids, contributing to the characteristic flavours and tastes of the various fermented products (Fox, 2009). The microbial community of African dairy-based fermented products predominantly comprises lactic acid bacteria (LAB), including species such as *Lactobacillus*, *Lactococcus* and *Streptococcus* species. While bacteria are the primary fermenters, yeasts such as *Saccharomyces* and *Candida* species also contribute to fermentation (Akabanda et al., 2013). In various African fermented milk products such as *leben* from Tunisia, *kivuguto* from Rwanda, *Masai* from Tanzania, *fènè* from Mali and *nunu* from Ghana, LAB counts ranging between 7.0 and 10.0 log CFU/mL have been observed (Akabanda et al., 2013; Bayili et al., 2019; Karenzi et al., 2013; Obodai & Dodd, 2006; Samet-Bali et al., 2012; Wullschleger et al., 2013). Commonly identified LAB species in these products include *Lactiplantibacillus plantarum*, *Lb. delbrueckii* subsp. *lactis*, *Lb. helveticus*, *Lacticaseibacillus casei* subsp. *casei*, *Lactococcus lactis*, *Leuconostoc mesenteroides*, *Leuconostoc pseudomesenteroides* among others (Ghosh et al, 2019). *Enterococcus* species such as *Enterococcus faecium* and *Enterococcus faecalis* have also been reported in some products, while *Streptococcus* species, particularly *Streptococcus infantarius* subsp. *infantarius* (*Sii*), have been frequently identified in certain traditional fermented milk products, indicating potential contamination from animal or human sources (Jans et al., 2012; Walsh et al., 2017; Wullschleger et al., 2013). Despite being part of intestinal microflora, certain strains of these species are used as probiotics (Agyei et al., 2020). The presence of particular LAB species like *Lp. plantarum* in dairy products, typically associated with African cereal-based fermented products such as *ogi* and *kenkey*, suggests possible contamination (Agyei et al., 2020; Wullschleger et al., 2013). Although less frequently documented in the literature, yeasts have also found in African fermented milk products, with notable significance in products like *amasi* from Zimbabwe, where yeast counts can reach up to 8.1 Log CFU/g (Gadaga et al., 2000). *Saccharomyces cerevisiae* is the predominant yeast species found in African fermented milk, often coexisting with *Candida* species and *Pichia kudriavzevii* (Akabanda et al., 2013; Bayili et al., 2019; Jespersen, 2003; Obodai & Dodd, 2006).

The LAB are a fermenting group of bacteria which play a major role in the production of African dairy-based fermented products, with occasional yeast identification towards the end of fermentation. The main microorganisms in various African fermented dairy products are listed in Table 4.2. The characterisation of LAB and yeasts in African

Table 4.2 Major Fermenting Microorganisms of African Dairy-Based Fermented Products

PRODUCT TYPE	PRODUCT NAME (COUNTRY/REGION)	ANIMAL ORIGIN	MICROBES RESPONSIBLE FOR FERMENTATION	REFERENCES
Fermented milk (Sour milk or yoghurt-like)	*Amasi* (South Africa; Zimbabwe)	Cow	*Lactiplantibacillus plantarum, Lactobacillus delbrueckii* subsp. *Lactis, Lactobacillus helveticus, Lactocaseibacillus casei* subsp. *casei, Lactocaseibacillus casei* subsp. *Pseudoplantarum, Lactococcus lactis, Lactococcus diacetylactis, Lactobacillus aeidophilus, Leueconostoc mesenteroides, Enterococcus aecium, Enterococcus aeealis, Saccharomyces cerevisiae, Saccharomyces dairenensis, Candida lusitaniae, Candida colliculosa, Dekera bruxillensis, Candida lipolytica, Candida tropicalis*	Feresu and Muzondo (1990), Gadaga et al. (2000), Todorov et al. (2007), Simatende et al. (2019), Maleke et al. (2021)
	Amabere (Kenya)	Cow	*Streptococcus thermophilus, Lactiplantibacillus plantarum, Leuconostoc mesenteroides, Saccharomyces cerevisiae and Candida famata*	Nyambane et al. (2014), Hawaz et al. (2025)
	Dhanaan (Ethiopia)	Camel	*Streptococcus* spp., *Lactococcus* spp., *Weissella* spp., *Streptococcus lutetiensis, Streptococcus infantarius.*	Berhe et al. (2019)
	Ergo (Ethopia)	Cow, goat and sheep	*Lactococcus garvieae, Lactococcus lactis* subsp. *lactis Lactiplantibacillus plantarum, Lactococcus lactic* subsp. *cremoris, Lactobacillus acidophilus, Leuconostoc lactis, Pediococcus pentosace, Lactobacillus* spp., *Streptococcus* spp., *Lactococcus* spp., *Leuconostoc* spp., *Entrococcus* spp., *and Pediococcus* spp. *Streptococcus thermophilus, Strep. acidominus, Enterococcus faecalis var. liquefaciens, Strep. bovis, Strep. mitis, Strep. agalactiae, Lactococcus cremoris, Leuconostoc dextranicum, Lactobacillus xylosus.*	Ashenafi (1995), Gonfa et al. (1999), Gonfa et al. (2001), Assefa et al. (2008)

(Continued)

Table 4.2 Major Fermenting Microorganisms of African Dairy-Based Fermented Products *(Continued)*

PRODUCT TYPE	PRODUCT NAME (COUNTRY/REGION)	ANIMAL ORIGIN	MICROBES RESPONSIBLE FOR FERMENTATION	REFERENCES
	Fene (Mali)	Cow	*Enterococcus* spp., *Limosilactobacillus fermentum, Lactiplantibacillus plantarum, Lactococcus lactis* subsp. *lactis, Pediococcus pentosaceus, Weissella confusa, Streptococcus infantarius* subsp. *infantarius, Streptococcus thermophilus*	Wullschleger et al. (2013), Agyei et al. (2020)
	Gariss (Sudan)	Camel	*Lactocaseibacillus paracasei* subsp. *paracasei, Limosilactobacillus fermentum, Lactiplantibacillus plantarum, Lactococcus lactis, Enterococcus* spp., *Leuconostoc* spp., *Streptococcus infantarius* subsp. *Infantarius, Streptococcus lactis* subsp. *diacetylactis, Lactocaseibacillus casei, Levilactobacillus brevis, Lactobacillus leichmanii, Lactobacillus acidophilus, Lactobacillus animalis, Lactobacillus divergens, Lactocaseibacillus rhamnosus, Lactobacillus gasseri, La. paracasei, Lactobacillus alimentarium, Lactobacillus* spp., *Lactococuss* spp.	Sulieman et al. (2006), Abdelgadir et al. (2008), Shori (2012)
	Kefir (South Africa)	Cow	*Lactobacillus delbrueckii* ssp. *delbrueckii, Lb. delbrueckii* ssp. *Lactis, Lb. curvatus, Lm. fermentum, Lactococcus lactis* ssp. *lactis, Leuconostoc lactis, Leu. mesenteroides* ssp. *Mesenteroides/dextranicum, Kluyveromyces marxianus, Saccharomyces cerevisiae, Kluyveromyces lactis, Saccharomyces unisporus, Saccharomyces rouxii, Torulaspora delbrueckii, Debaryomyces hansenii, Candida kefyr, C. lipolytica, C. holmii, Zygosaccharomyces* spp.	Loretan et al. (2003), Witthuhn et al. (2004)
	Kindirmo (Nigeria)	Cow	*Lactobacillus hilgardi, Lactocaseibacillus casei, Lactiplantibacillus plantarum, Lactobacillus bifidum, Lactobacillus lactis and Lactobacillus bulgaricus*	Igwe et al. (2014), Bamgbose et al. (2022)

Kivuguto (Rwanda)	Cow	*Lactococcus lactis, Leuconostoc mesenteroides* subsp. *mesenteroides, Leuconostoc pseudomesenteroides*	Karenzi et al. (2013), Agyei et al. (2020)
Kule-Naoto (Kenya)	Cow	*Lactiplantibacillus plantarum, Limosilactobacillus fermentum, Lactocaseibacillus paracasei, Lactobacillus acidophilus*	Mathara et al. (2004)
Kwerionik (Uganda)	Cow	*Lactiplantibacillus plantarum, Enterococcus faecalis, Lactocaseibacillus paracasei* subsp. *paracasei, Lactocaseibacillus casei* subsp. *casei, Lactococcus lactis* subsp. *lactis, Enterococcus faecium and Leuconostoc mesenteroides* subsp. *mesenteroides*	Nakavuma et al. (2012), Agyei et al. (2020)
Lait caille (Burkina Faso, Côte d'Ivoire, Senegal)	Cow	*Leuconostoc mesenteroides, Pediococcus pentosaceus, Weissella paramesenteroides, Lactococcus lactis, Enterococcus* spp., *Candida parapsilosis and Saccharomyces cerevisiae*	Bayili et al. (2019)
Laban/Leben/Lben (Tunisia, Morocco, Libya, Egypt)	Cow	*Lactococcus lactis* subsp. *lactis, Lactococcus lactis* subsp. *cremoris, Leuconostoc mesenteroides, Enterococcus faecalis, E. faecium, E. hirae, E. durans, E. casselifavus, E. saccharolyticus, Pediococcus pentosaceus, Lactococcus lactis, Lactiplantibacillus plantarum, La. rhamnosus, La. casei, Lv. brevis, Lb. acidophillus, Lm. fermentum, Candida krusei, Candida tropicalis, Candida lusitania*	Samet-Bali et al. (2012), Saleh (2013), Samet-Bali et al. (2016), Sarhir et al. (2019), Alnakip et al. (2023), Nahaisi et al. (2024)
Mabisi (Zambia)	Cow	*Lactococcus lactis, Lactocaseibacillus casei, Lactobacillus helveticus, Lysinibacillus sphaericus, Streptococcus thermophilus, Streptococcus equinus, Acetobacter pasteurianus, Acinetobacter* spp., *Acinetobacter ursingii, Chryseobacterium bovis, Leuconostoc garlicum, Leuconostoc pseudomesenteroides, Enterococcus durans, Levilactobacillus brevis, Lactiplantibacillus plantarum, Lactobacillus kefiranofaciens* subsp. *kefirgranum,*	Schoustra et al. (2013), Agyei et al. (2020), Moonga et al. (2020)

(Continued)

Table 4.2 Major Fermenting Microorganisms of African Dairy-Based Fermented Products *(Continued)*

PRODUCT TYPE	PRODUCT NAME (COUNTRY/REGION)	ANIMAL ORIGIN	MICROBES RESPONSIBLE FOR FERMENTATION	REFERENCES
	Madila (Botswana)	Goat	*Lactobacillus acidophilus*, *Lactobacillus delbrueckii*, *Lactiplantibacillus plantarum*, *Lactococcus lactis*, *Limosilactobacillus fermentum*, *Levilactobacillus brevis*	Schutte (2013), Agyei et al. (2020)
	Masse (Mozambique)	Cow	*Lactococcus lactis* subsp. *lactis*, *Leuconostoc pseudomesenteroides*, *Leuconostoc lactis*, *Leuconostoc garlicum*, *Lactococcus* spp.	Schutte (2013), Agyei et al. (2020)
	Mursik (Kenya)	Cow or goat	*Lactocaseibacillus casei*, *Lactococcus casei* subsp. *casei*, *Lactobacillus confusus*, *Lactobacillus delbruenkii bulgaricus*, *Lactobacillus lactis*, *Lactiplantibacillus plantarum*, *Leuconostoc mesenteroides*, *Enterococcus faecium*, *Streptococcus salivarius*, *Streptococcus diacetylactis*, *Candida krusei*, *Candida kefyr*, *Candida sphaerica*, *Saccharomyces fermentati*	Nieminen et al. (2013), Digo et al. (2017)
	Nunu (Ghana, Nigeria)	Cow	*Streptococcus infantarius*, *Limosilactobacillus fermentum*, *Lactiplantibacillus plantarum*, *Leuconostoc mesenteroides*, *Lactobacillus helveticus*, *Enterococcus faecium*, *Enterococcus italicus*, *Weissella confusa*, *Candida parapsilosis*, *Candida rugosa*, *Candida tropicalis*, *Galactomyces geotrichum*, *Pichia kudriavzevii*, *Saccharomyces cerevisiae*, *Saccharomyces pastorianus*, *Candida kefyr*, *Yarrowia lipolytica*, *Candida stellata*, *Kluyveromyces maxianus*	Akabanda et al. (2010), Akabanda et al. (2013), Walsh et al. (2017)

Nyarmie (Ghana)	Cow	*Leuconostoc mesenteroides* subsp. *mesenteroides, Streptococcus thermophilus, Lactobacillus delbrueckii* subsp. *bulgaricus, Lactobacillus helveticus, Lactobacillus delbrueckii* subsp. *Lactis, Lactococcus lactis, Saccharomyces cerevisiae, Candida* spp.	Obodai and Dodd (2006)
Pendidaam (Cameroon)	Cow	*Lactobacillus delbrueckii, Lactobacillus helveticus, Limosilactobacillus fermentum, Lactiplantibacillus plantarum, Lactocaseibacillus casei, Streptococcaceae, Enterococcus faecalis var. faecalis and Enterococcus faecium, Streptococcus salivarius var. thermophilus, Leuconostoc mesenteroides*	Jiwoua and Milliére (1990), Mbawala et al. (2013), Agyei et al. (2020)
Roub/*Rob* (Sudan)	Cow	*Limosilactobacillus fermentum, Lactobacillus acidophilus, Lactococcus lactis, Streptococcus salivarius, Saccharomyces cerevisiae, Candida kefyr*	Abdelgadir et al. (1998), Abdelgadir et al. (2001), Agyei et al. (2020)
Suusac (Kenya, Somali)	Camel	*Lactobacillus curvatus, Lactiplantibacillus plantarum, Ligilactobacillus salivarius, Lactococcus raffinolactis, Leuconostoc mesenteroides* subsp. *mesenteroides, Streptococcus infantarius* subsp. *infantarius* and *Streptococcus gallolyticus* subsp. *macedonicus, Candida krusei, Geotrichum penicillatum, Rhodotorula mucilaginosa*	Lore et al. (2005), Shori (2012), Jans et al. (2012)
Zabady (Egypt)	Buffalo, cow, or a mixture of both	*Streptococcus thermophilus, Lactococcus garvieae, Lactococcus raffinolactis, Lactococcus lactis, Leuconostoc citreum, Lactobacillus delbrueckii* subsp. *bulgaricus and Lactobacillus johnsonii, Enterococcus faecalis, E. faecium, E. hirae, E. durans, Lactiplantibacillus plantarum, La. rhamnosus, La. casei, Lv. brevis, Lb. acidophillus, Lm. fermentum, Leuconostoc mesenteroides*	El-Baradei et al. (2008), Labtar et al. (2019), Alnakip et al. (2023)

(Continued)

Table 4.2 Major Fermenting Microorganisms of African Dairy-Based Fermented Products *(Continued)*

PRODUCT TYPE	PRODUCT NAME (COUNTRY/REGION)	ANIMAL ORIGIN	MICROBES RESPONSIBLE FOR FERMENTATION	REFERENCES
Cheese	*Ayib* (Ethopia)	Cow	*Lactobacillus fermenti*, *Lactiplantibacillus plantarum*, *Lactocaseibacillus casei*, *Lactobacillus acidophilus*, *Brevibacterium* spp., *Kluyveromyces* spp., *Sporobolomyces* spp., *Candida* spp., *Torulopsis*, *Leucosporidium*, *Kluyveromyces lactis*, *Kluyveromyces bulgaricus*, *Candida pseudotropicalis*	Ashenafi (1989), Ashenafi (1994)
	Jben (Morocco)	Cow	*Lactiplantibacillus plantarum*, *Lactocaseibacillus rhamnosus*, *Lactocaseibacillus paracasei*, *Levilactobacillus brevis*, *Lactobacillus buchneri*, *Lactococcus lactis*, *Lactococcus garvieae*, *Lactococcus raffinolactis*, *Leuconostoc pseudomesenteroides*, *Leuconostoc mesenteroides*, *Leuconostoc citreum*, *Enterococcus durans*, *Enterococcus faecium*, *Enterococcus gilvus*, *Enterococcus hirae*, *Streptococcus* sp.	Ouadghiri et al. (2009)
	Klila (Algeria)	Cow	*Lactiplantibacillus plantarum*, *Lactocaseibacillus casei*, *Lactobacillus acidophilus*, *Levilactobacillus brevis*, *Lactobacillus alimentarius*, *Lactobacillus intestinalis*, *Lactobacillus helveticus*, *Enterococcus faecalis*, *Enterococcus faecium*, *Lactobacillus confuses*, *Streptococcus* spp., *Pediococcus* spp., *Pediococcus acidilactici and Leuconostoc* spp.	Boubekri and Ohta (1996), Guetouache and Guessas (2015)
	Kariesh (Egypt)	Cow	*Enterococcus faecalis*, *E. faecium*, *E. hirae*, *E. durans*, *E. casseliflavus*, *E. saccharolyticus*, *Pediococcus pentosaceus*, *Lactococcus garvieae*, *Lc. lactis*, *Lactiplantibacillus plantarum*, *La. rhamnosus*, *La. casei*, *Lv. brevis*, *Lb. acidophillus*, *Lm. fermentum*, *Leuconostoc mesenteroides*	Alnakip et al. (2023)

	Warasanki/Wara (Nigeria)	Cow	*Lactobacillus* spp., *Lactococcus* ssp., *Leuconostoc* spp., *Pediococcus* spp., *Lactiplantibacillus plantarum and Lactobacillus acidophilus*	Sangoyomi et al. (2010), Ajibola et al. (2020)
Butter	*Smen* (Algeria, Morroco)	Cow, Goat, Ewe	*Lactiplantibacillus plantarum*, *Lactocaseibacillus casei*, *La. rhamnosus*, *Levilactobacillus brevis*, *La. paracasei*, *Lb. parabuchneri*, *Lb. mesenteroides*, *Lb. lactis*, *Lb. garvieae*, *Leuconostoc mesenteroides*, *E. faecium*, *E. durans*, *Enterococcus faecalis*, *Clavispora lusitaniae*, *Rhodotorula mucilaginosa*, *W. pararugosa*, *Kluyveromyces marxianus*, *Saccharomyces cerevisiae/paradoxus*, *Y. lipolytica*, *Candida sorbosivorans*, *Torulaspora delbrueckii*, *K. lactis*, *G. candidum*, *M. dehoogii*, *Moniliella suaveolens*, *Moniliella species* Early maturation stage: *Bacillus alvei*, *Bacillus firmus*, *Bacillus brevis*, *Bacillus cereus*	Tantaoui-Elaraki and El Marrakchi (1987), Labtar et al. (2019), Boussekine et al. (2022)
	Zebda beldia (Northern African)		*Lactobacillus lactis* subsp. *Lactis*, *Lactobacillus* subsp. *Cremoris*, *Streptococcus salivarius* subsp. *thermophilus*, *Lactobacillus delbrueckii* subsp. *bulgaricus*, *Lactiplantibacillus plantarum*, *yeasts: Saccharomyces cerevisiae and Kluyveromyces marxianus.*	Benkerroum (2013)
Ghee	*Mashita* (Uganda)	Cow	*Lactocaseibacillus paracasei*, *Lactobacillus helveticus*, *Lactiplantibacillus plantarum*, *Lactobacillus perolens*, *Enterococcus faecium*, *Levilactobacillus brevis*, *Lactobacillus acetotolerans*, *Bifidobacterium* sp., *Lactobacillus* sp., *Lactococcus raffinolactis*, *Lactococcus lactis* subsp. *lactis*, *Streptococcus salivarius.*	Ongol and Asano (2009)

(Continued)

Table 4.2 Major Fermenting Microorganisms of African Dairy-Based Fermented Products *(Continued)*

PRODUCT TYPE	PRODUCT NAME (COUNTRY/REGION)	ANIMAL ORIGIN	MICROBES RESPONSIBLE FOR FERMENTATION	REFERENCES
Buttermilk	*Omashikawa* (Namibia)	Cow	*Lactobacillus helveticus, Lactobacillus kefiri, Lactobacillus casei, Lactocaseibacillus rhamnosus, Lactocaseibacillus paracasei, Lactococcus lactis* subsp. *Lactis, Leuconostoc pseudomesenteroides, Enterococcus faecium, Enterococcus durans*	Schutte (2013), Agyei et al. (2020)
	Hazo (Ethiopia)	Cow	*Lactococcus lactis* ssp. *lactis, Lactococcus lactis* spp. *cremoris, Enterococcus faecalis, Lactiplantibacillus plantarum, Lactobacillus pentosus, Lactocaseibacillus paracasei* subsp. *casei, Levilactobacillus brevis, Leuconostoc mesenteroides, Saccharomyces cerevisiae, Kluyveromyces marxianus, Candida parapsilosis, Trichosporon asahii*	Gebreselassie et al. (2016)

Abbreviations: La., Lacticaseibacillus; Lb., Lactobacillus; Li., Ligilactobacillus; Lm., Limosilactobacillus; Lp., Lactiplantibacillus; Lv., Levilactobacillus.

spontaneous fermented milk products has primarily been conducted on end-products, relying on phenotypic techniques which have limited discriminatory power (Akabanda et al., 2013; Gadaga et al., 2000; Savadogo et al., 2004). Studies on microbial successions while processing African fermented milk products remain relatively limited (Akabanda et al., 2013; Wullschleger et al., 2013). This chapter elaborates on the processing and microbiology of African fermented dairy products from various regions of Africa.

4.2 Selected Fermented Milk Products from Southern Africa

4.2.1 Amasi

Amasi, also known as *maas*, is a traditional fermented dairy product enjoyed with maize cereal (crumbled pap) in Southern Africa (Maleke et al., 2021). The nutritional composition of amasi varies widely, influenced by factors such as milk composition, fermentation temperature, duration and storage conditions (Maleke et al., 2021; Moonga et al., 2019).

The production of *amasi* involves a straightforward yet intricate process. Raw cow milk is collected and left to ferment naturally at room temperature (around 25°C) for approximately 24–48 hours (Figure 4.1). Primarily *Lactococcus lactis* subsp. *lactis*, *Lactococcus lactis* subsp. *cremoris*, *Lactobacillus* species or *Streptococcus* species, present in the fresh milk, ferment lactose into lactic acid to produce *amasi* (Beukes et al., 2001; Narvhus & Gadaga, 2003). As fermentation progresses, other LAB species may also contribute to the microbial community. These LAB lower the milk's pH, creating an acidic environment that inhibits the growth of spoilage organisms. Additionally, LAB-produced proteolytic enzymes break down milk proteins into smaller peptides and amino acids, enhancing amasi's characteristic flavour and texture, resulting in its tangy flavour and thickened consistency (Beukes et al., 2001). The predominant yeast species present during fermentation are *Saccharomyces cerevisiae*, *S. dairenensis*, *Candida lusitaniae* and *C. colliculosa* (Gadaga et al., 2000). The fermentation process typically reduces the lactose content in *amasi* in comparison to fresh milk, making it suitable for lactose-intolerant individuals (Du Plooy et al., 2018). Once fermentation reaches the

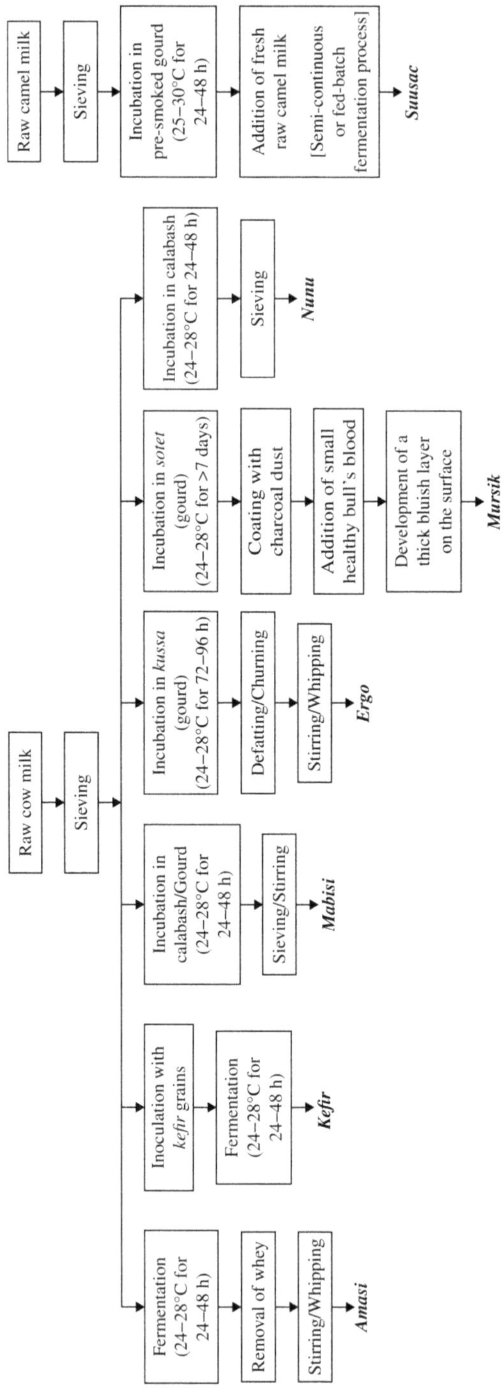

Figure 4.1 Schematic flow chart for the production of various traditional African fermented milk products by spontaneous fermentation.

desired level, the product is cooled to halt bacterial activity and may be refrigerated for storage or consumed fresh. *Amasi* is renowned for its probiotic properties and nutritional benefits, making it a popular choice in Southern African cuisine. In addition, the low pH resulting from organic acid production during fermentation aids in delaying gastric emptying when *amasi* is consumed, which has positive effects on glycemic response, satiety and appetite control (Du Plooy et al., 2018; Vorster et al., 2013).

4.2.2 Kefir

Kefir is a unique cultured milk dairy product consumed around the world, which combines lactic acid with the alcoholic fermentation of lactose. *Kefir* is a fermented milk beverage with an acidic flavour and is thick and slightly bubbly (Plessas et al., 2016). *Kefir* differs from other fermented products due to its unique its starter, which consists of *kefir* grains (Bengoa et al., 2019). The *kefir* grains consist of a symbiotic matrix of LAB, yeasts and polysaccharides (Loretan et al., 2003). In South Africa, naturally fermented *kefir* is often of inconsistent quality due to the bacteria and yeasts present in the *kefir* grains, which are known to vary widely from those in other parts of the world (Witthuhn et al., 2004). The microbiology of *kefir* from this region reflects a diverse and dynamic microbial community, which contributes to its distinctive properties. The fermentation process of *kefir* in South Africa involves inoculating milk with kefir grains, followed by incubation at ambient temperatures for 24–48 hours (Figure 4.1). During fermentation, LAB and yeasts work synergistically, with the LAB converting lactose into lactic acid, lowering the pH and causing milk proteins to coagulate, resulting in a thicker consistency. During this process the yeasts convert lactose to ethanol and CO_2, contributing to the slight alcoholic content of *kefir* (usually less than 1%) and its carbonation (Plessas et al., 2016).

The key microorganisms involved in South African *kefir* fermentation include various species of LAB, predominantly *Lactobacillus delbrueckii* subsp. *delbrueckii*, *Lb. delbrueckii* subsp. *lactis* and *Leuconostoc lactis* (Witthuhn et al., 2004), and yeasts, mainly *Saccharomyces cerevisiae*, *Kluyveromyces marxianus*, *Kluyveromyces lactis*, *Candida lipolytica* and *C. kefyr* (Loretan et al., 2003; Witthuhn et al., 2004). The bacteria play a crucial role in fermenting lactose into lactic acid, which lowers the pH of the milk, inhibits pathogenic microorganisms and contributes to the tangy flavour of *kefir*. The *Leuconostoc lactis* and *Leuconostoc mesenteroides* are also known for their role in producing the gas CO_2, contributing to the effervescence of *kefir* (Plessas et al., 2016). Yeasts are equally important in kefir fermentation, contributing to ethanol and carbon dioxide production, which add to the effervescence and flavour complexity.

4.2.3 Mabisi

Mabisi, also known as *mabila*, is a traditional fermented milk product from Zambia. It is a staple in many Zambian households and is appreciated for its unique taste, texture and nutritional benefits. *Mabila* production begins with raw cow milk collection, traditionally collected in a calabash (gourd) or earthen pots but plastic or metal containers can also be used due to cost and availability (Groenenboom et al., 2020). Households tend to have their own calabashes, which they use for consecutive rounds of *Mabisi* production. The methods of producing *Mabisi* vary depending on location and can include back-slopping and churning, among other methods (Groenenboom et al., 2020; Moonga et al., 2019). The milk is left to ferment naturally at ambient temperatures, typically for one to three days (Figure 4.1), during which time naturally occurring microorganisms in the milk, as well as those introduced from the environment and utensils used, aid in the fermentation, after which the fermented product is stirred before consumption (Moonga et al., 2019). The microbiology of *mabila* is diverse, with LAB playing a crucial role in the fermentation. A report identified six to ten different species of LAB that dominated fermentation (Schoustra et al., 2013).

The prominent bacteria found in *mabila* is *Lactococcus lactis*, one of the primary LAB dominating the fermentation process (Groenenboom et al., 2020; Moonga et al., 2020). Other LAB such as *Lactiplantibacillus plantarum*, *Lb. kefiranofaciens*, *Lb. helveticus* and *Streptococcus thermophiles* are also commonly found in *mabila*. These bacteria contribute to the development of the characteristic tangy flavour and texture of *mabila* through their metabolic activities.

4.3 Selected Fermented Milk Products from East Africa

4.3.1 Suusac

In East Africa, camels play an important role in food security. Fermented camel milk (*suusac*) is an integral part of the nomad's diet in East Africa, in the pastoral communities in Kenya and Somalia (Farah et al., 2007; Jans et al., 2012). The traditional production and fermentation of *suusac* begin with fresh camel milk being spontaneously fermented in a gourd pre-smoked using flaming acacia twigs or African wild olive leaves (*Acacia busia* or *Olea Africana*) and incubated at ambient temperature (25–30°C) for one to two days (Lore et al., 2005). This process results in a microflora primarily composed of LAB, likely originating from the vessel's surface (Lore et al., 2005). The occasional addition of unfermented (raw) camel milk to already fermented *suusac* could serve as another source of typical dairy for the LAB species for further fermentation (Jans et al., 2012; Shori, 2012). Hence, *suusac* production is a semi-continuous or fed-batch fermentation process. The microbiology of *suusac* typically harbours LAB counts ranging from 7.0 to 8.0 log CFU/mL (Jans et al., 2012). The LAB species identified in *suusac* include *Lactococcus lactis*, *Limosilactobacillus fermentum*, *Lactiplantibacillus plantarum*, *Pediococcus acidilactici*, *Streptococcus thermophilus* and *Streptococcus infantarius* subsp. *infantarius* (*Sii*) (Jans et al., 2012). A report by Jans et al. (2012) confirmed that *suusac* contained predominantly *Streptococcus infantarius* subsp. *infantarius* followed by *Lactococcus lactis* subsp. *lactis*, *Streptococcus thermophilus* and *lactobacilli* species. These LAB are responsible for the characteristic tangy flavour and texture of the *suusac*. In addition, yeast species such as *Candida krusei* and *Rhodotorula mucilaginosa* have been detected in suusac, although their prevalence and contribution to fermentation dynamics require further investigation (Fugl et al., 2017). The microbial succession during suusac fermentation is influenced by various factors such as milk composition, temperature and the use of traditional fermentation techniques involving wooden containers and leaves for packaging (Fugl et al., 2017).

4.3.2 Ergo

Ergo, also known as Ethiopian fermented milk, is a traditional dairy product widely consumed in Ethiopia. It is similar to yoghurt but has distinct microbial and sensory characteristics due to the unique fermentation process and local microbial flora involved (Assefa et al., 2008). Traditionally, production of *ergo* involves raw (unpasteurized) cow's milk collected in a *kussa* (gourd) and left to ferment naturally at ambient temperatures for three to four days, allowing the native microbial flora to initiate fermentation (Gonfa et al., 1999, 2001). The LAB and yeasts grow and multiply during fermentation and interact to produce a fermented product with a low pH (typically around 4.0–4.5), thick texture and distinct sour taste. A study of *ergo* fermented in pre-smoked traditional container (*kussa*) indicated that LAB predominated with counts greater than Log_{10} 8 CFU/mL, followed by yeasts with counts Log_{10} 5 CFU/mL and moulds Log_{10} 4 CFU/mL (Gonfa et al., 2001).

The microbiology of *ergo* is crucial to understanding its production, safety and nutritional benefits. The primary microorganisms involved in the fermentation of *ergo* are LAB and yeasts. The *Lactococcus garvieae* and *Lactococcus lactis* subsp. *lactis* are the dominant strains of LAB typically found in *ergo* (Assefa et al., 2008; Gonfa et al., 2001). Other LAB include *Lactobacillus* species such as *Lactiplantibacillus plantarum*, *Lactobacillus delbrueckii* subsp. *bulgaricus* and *Limosilactobacillus fermentum* are commonly found in *ergo*, as well as *Streptococcus thermophilus*, *Strep. acidominus*, *Strep. bovis*, *Strep. mitis*, *Strep. agalactiae*, *Enterococcus faecalis* var. *liquefaciens*, *Lactococcus cremoris*, *Leuconostoc dextranicum*, *Leuc. lactis*, *Lactobacillus xylosus* and *Lact. lactis* (Gonfa et al., 2001). The presence of LAB in *ergo* not only contributes to its sensory properties but also offers potential health benefits. LAB are known for their probiotic effects, which include improving gut health, enhancing the immune system and inhibiting the growth of pathogenic bacteria. Some LAB strains from isolated ergo have been proven to have some antimicrobial activity against pathogenic microorganisms (Lelise et al., 2014). The production of organic acids and bacteriocins by LAB helps preserve the product by preventing spoilage and the growth of harmful microorganisms. However, the traditional non-pasteurised *ergo* production process raises concerns about the potential presence of pathogenic bacteria.

4.3.3 Mursik

Mursik, a traditional fermented cow's or goat's milk beverage consumed by the Kalenjin community of Kenya, has a rich microbiological profile which contributes to its unique taste, texture and nutritional properties. The production of *mursik* involves using specific calabash gourds called *sotet*. Several days before fresh milk collection, a branch from the Iteet tree (*Senna didymobotrya*) is stripped of its bark and dried. The end of the stick is charred in a fire and rubbed repeatedly on the cleaned gourd's interior until it is fully coated with charcoal dust (Digo et al., 2017; Kirui & Nguka, 2017; Nieminen et al., 2013). This process decreases the gourd's porosity and enhances the flavour of the milk. Charcoal treatment of the fermentation gourds may modify the odour, taste and flavour of *mursik* by decreasing the milk's pH and impacting the growth and metabolism of the microbes. Some Kalenjin groups also add small amounts of blood from a healthy bull, which is collected by pricking a vein in the neck and removing the fibrin through gentle stirring (Kirui & Nguka, 2017; Nieminen et al., 2013). This addition can influence microbial metabolism, as iron is a crucial cofactor for various cellular processes. When *mursik* is ready, it develops a thick bluish layer on the surface, which is shaken well before consumption to form a uniform emulsion (Kirui & Nguka, 2017; Nieminen et al., 2013). In many Kalenjin households, fermented milk is consumed multiple times daily. The LAB group, particularly species of *Lactobacillus*, *Leuconostoc* and *Streptococcus* species, are the primary microorganisms involved in the fermentation of *mursik* (Digo et al., 2017). The *Lactobacillus kefiri* is the predominant LAB strain which has been isolated from *mursik*, and other common LAB are *Lactocaseibacillus casei*, *La. paracasei* and *La. rhamnosus* (Nieminen et al., 2013). The coexistence of LAB and yeasts has been reported in *mursik* fermentation, with *Candida krusei* being the most commonly identified yeast, while *Candida kefyr*, *Candida sphaerica* and *Saccharomyces fermentati* have also been identified (Nieminen et al., 2013).

4.4 Selected Fermented Milk Products from North Africa

4.4.1 Zabady

Traditional *zabady* is Egypt's most widely-enjoyed fermented milk, known for its cooked flavour and yoghurt-like firmness (El-Baradei et al., 2008). It is believed to be one of the oldest fermented milk products in the world, particularly in the Middle East (El-Baradei et al., 2008; El-Gendy, 1983). The highest quality *zabady* is primarily made from buffalo milk, though it can also be made from cow's milk or a mix of both (Abou-Donia, 2004). The traditional preparation begins by boiling raw milk briefly, then cooling it to around 45°C (Figure 4.1). Backslopping is done by adding a portion of the previous day's *zabady* as a starter, which contains mainly *Streptococcus thermophilus* and *Lactobacillus delbrueckii* subsp. *bulgaricus*. The inoculated milk is then poured into porcelain pots or plastic cups and incubated at approximately 42°C until coagulation, after which it is cooled to prevent further acidity development (El-Baradei et al., 2008).

The LAB identified in fermented *zabady* are mostly *Streptococcus thermophilus*, *Lactococcus garvieae*, *Lactococcus raffinolactis*, *Lactococcus lactis*, *Leuconostoc citreum*, *Lactobacillus delbrueckii* subsp. *bulgaricus* and *Lactobacillus johnsonii* (El-Baradei et al., 2008), as well as *Enterococcus faecalis*, *E. faecium*, *E. hirae*, *E. durans*, *Lactiplantibacillus plantarum*, *La. rhamnosus*, *La. casei*, *Lv. brevis*, *Lb. acidophilus*, *Lm. fermentum* and *Leuconostoc mesenteroides* (Alnakip et al., 2023). The dominant LAB strains in the fermentation are *Streptococcus thermophilus* and *Lactococcus garvieae*. The streptococcal and lactococcal groups are the major bacterial species in *zabady* fermented milk, while the *lactobacilli* appear in lessor quanties (El-Baradei et al., 2008). The known probiotic *Lactobacillus johnsonii* has been identified in *zabady* (El-Baradei et al., 2008).

4.4.2 Smen

Smen, also known as *semna* (Egypt), *samin* (Sudan), *smen* (Morocco), and *smen* or *dhan* (Algeria), is a traditional fermented butter made from raw whole milk derived from cows, goats or ewes, either in whole or as a blend (Boussekine et al., 2020, 2022). This butter is classified as rancid butter made from salted raw butter (containing 8–10% salt) derived from spontaneously acidified milk (*raib*). It is aged in the dark under cool, anaerobic conditions (13–15°C) for 1–12 months or even longer (Boussekine et al., 2022; Iradukunda et al., 2018). It is important to note that the preparation method of *smen* varies across countries, and knowledge about its preparation has been passed down through generations.

The preparation of this butter uses raw milk which is spontaneously fermented at ambient temperature for 12–120 hours. The process includes coagulation of the fermented milk, and the coagulum called *raib* is churned with a traditional churn to separate the butterfat from the buttermilk. During churning, warm water or milk is added to the mixture to encourage the agglomeration of butter grains. After churning, the butter (*zebda*) is recovered using a perforated ladle. The butter is washed several times to remove the buttermilk and traces of impurity (Figure 4.2). The washed butter is mixed with a significant amount of salt and sometimes herbs, homogenised and covered for one week to remove water, then packaged tightly in a container, usually made of clay, and buried or stored in a cool, dark place for several months to years (Boussekine et al., 2022).

The microbiology of *smen* is a result of its fermentation and maturation stages. The primary groups of microorganisms involved are LAB, mainly the *Enterococci* and *Lactobacilli* species. *Enterococcus faecium* and *Enterococcus durans* are crucial in the fermentation process, while *Lentilactobacillus parabuchneri*, *Lacticaseibacillus paracasei*, *Lacticaseibacillus rhamnosus*, *Lactiplantibacillus plantarum* and *Latilactobacillus curvatus* occur in variable proportions during fermentation (Boussekine et al., 2022). Yeasts present at different stages of fermentation and maturation are *Clavispora lusitaniae*, *Rhodotorula mucilaginosa*, *W. pararugosa*, *Kluyveromyces marxianus*, *Saccharomyces cerevisiae/paradoxus*, *Y. lipolytica*, *Candida sorbosivorans*, *Torulaspora delbrueckii*, *K. lactis*, *G. candidum*, *M. dehoogii* and *Moniliella suaveolens* (Boussekine et al., 2022). The microbial succession in *smen* during maturation has been identified to be *La. paracasei*, *Lb. lactis*, *La. rhamnosus* and *Lb. garvieae*, which are dominant from initiation, after one-month *Lb. mesenteroides*, *Lb. lactis*, *La. rhamnosus* and *Lb. garvieae* can be identified, and at three-months only the *Lv. brevis*, *E. faecium* and *La. paracasei* species are present (Boussekine et al., 2022). *S. cerevisiae*, *K. marxianus*, *Y. lipolytica*, *C. sorbosivorans*, *W. pararugosa and T. delbrueckii* yeasts have been detected from initiation. After one month of maturation, *G. candidum* and *Moniliella* species can also be detected, while at three months, only *Moniliella* spp. and *C. sorbosivorans* can be detected (Boussekine et al., 2022).

Several biochemical changes occur in *smen* during production. For example, in early maturation stages, such as during lipid hydrolysis and protein breakdown, lipolytic enzymes break down triglycerides into free fatty acids, glycerol, and other compounds, which contribute

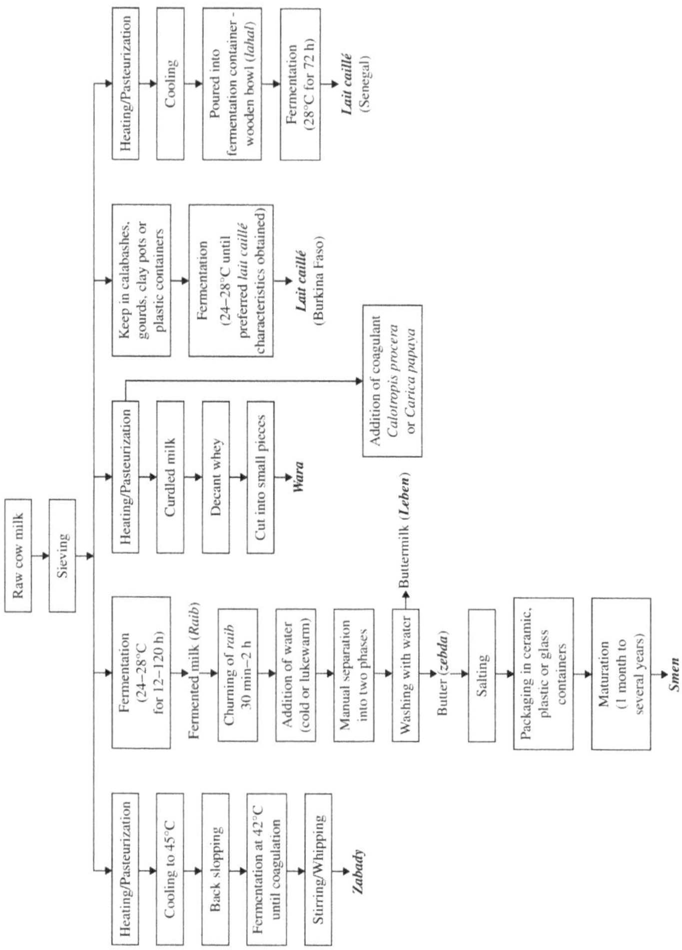

Figure 4.2 Schematic flow chart for the production of various traditional African fermented dairy products.

to the characteristic rancid flavour of *smen*. The proteolytic enzymes hydrolyse proteins into peptides and amino acids. This proteolysis can contribute to the umami flavour and further enhance the aroma. These are reported to be produced by *Bacillus alvei*, *Bacillus firmus*, *Bacillus brevis* and *Bacillus cereus* (Tantaoui-Elaraki & El Marrakchi, 1987). However, these *Bacillus* species tend to disappear as maturation progresses. Aromatic compound production is initiated by the LAB microbial metabolism, which leads to the formation of various aromatic compounds such as esters, ketones and aldehydes, which contribute to the complex flavour profile of *smen* during maturation (Boussekine et al., 2022).

4.4.3 Leben

Leben, known as *Lben* or *Laban*, is a traditional fermented buttermilk product widely consumed in Algeria, Egypt, Libya, Morocco and Tunisia. The production of traditional *leben* involves the spontaneous fermentation of cow's milk by a community of natural microflora, primarily LAB. The fermentation takes 18–24 hours, after which the fermented milk (*raib* or *rayeb*) is churned. Traditionally, churning takes place in a goat leather bag called *checoua*. The churning is achieved by hanging the *checoua* filled with *raib* and vigorously shaking it back and forth until the fat globules coalesce. Through churning, the fermentation is separated into aqueous fractions, producing *leben* and raw butter (*zebda beldi*), which is the fatty fraction (Samet-Bali et al., 2012, 2016). The schematic flowchart of *leben* production is depicted in Figure 4.2. The *checoua* is salted after two successive churning cycles to avoid an unpleasant odour.

Diversity of LAB identified in *leben* is listed in Table 4.2; this comprises mainly members of the *Lactobacilli*, *Lactococci* and *Leuconostocs* genera. The dominant LAB species is *Lactococcus lactis* subsp. *lactis* during *leben* production; however, *Lactococcus lactis* subsp. *cremoris* are frequently identified, while the most frequently identified yeast is *Candida krusei* (Bensalah et al., 2009; Samet-Bali et al., 2016). In Tunisian *leben*, *Lactobacillus* is not usually identified (Samet-Bali et al., 2016), although it may be detected in lower counts in Moroccan *leben* (Benkerroum & Tamime, 2004), while *Lactobacillus* species dominate the Libyan leben (Nahaisi et al., 2024). In *leben* fermentation, the yeast counts are usually lower relative to the LAB. Yeasts in traditional *leben* are recovered towards the end of the fermentation stage, which suggests that they play a secondary role in the fermentation process, particularly in the aroma development in the product (Samet-Bali et al., 2016).

4.5 Selected Fermented Milk Products from West Africa

4.5.1 Nunu

Nunu or *nono* is a traditional fermented milk product widely consumed in Ghana, Nigeria and other West African countries. *Nunu* is a yoghurt-like dairy product made solely from cow's milk. The production involves a spontaneous fermentation of raw milk (no pre-fermentation heating) in calabashes or plastic containers under ambient conditions for 24–48 hours, after which it can be consumed (Akabanda et al., 2010, 2013). The LAB which have been identified during fermentation include *Lactobacillus*, *Leuconostoc*, *Lactococcus*, *Enterococcus* and *Streptococcus* species, while yeasts reported are *Saccharomyces cerevisiae*, *Saccharomyces pastorianus*, *Candida kefyr*, *Yarrowia lipolytica*, *Candida stellata*, *Kluyveromyces maxianus*, *Zygosaccharomyces bisporus* and *Zygosaccharomyces rouxii* (Akabanda et al., 2010).

The diversity of LAB and yeast in *nunu* is reported in Table 4.2. The predominant LAB found from the initiation to the end of fermentation during *nunu* production is *Limosilactobacillus fermentum*, while *Lactiplantibacillus plantarum* and *Leuconostoc mesenteroides* play prominent roles during the first 6–8 hours of fermentation (Akabanda et al., 2013). Other LAB, such as *Lactobacillus helveticus*, *Enterococcus faecium*, *Enterococcus italicus*, *Weissella confusa* and *Lactococcus* spp., are less frequently isolated. Another study, using metagenomic analysis to investigate microbial succession in the *nunu* fermentation process, reported that the *Streptococcus infantarius* dominated, followed by *Lactococcus lactis* (Walsh et al., 2017). The yeasts *P. kudriavzevii* and *S. cerevisiae* are the dominant yeast species. *C. parapsilosis*, *Candida rugosa* and *C. tropicalis* have been reported sporadically during fermentation, with *G. geotrichum* contributing a substantial part of the yeasts in the later stages of the fermentation (Akabanda et al., 2013).

4.5.2 Lait Caillé

Lait caillé, also known as curdled milk, is a traditional fermented dairy product from West Africa, particularly Burkina Faso and Senegal. Typically, *lait caillé* is produced by the spontaneous fermentation of cow's milk, but there are some distinct differences in production steps between the two countries. In Burkina Faso, *lait caillé* is prepared from raw cow milk, or occasionally, from raw goat milk. The preparation of *lait caillé* starts by collecting the raw milk in calabashes, gourds, clay pots or plastic containers, and the milk is fermented at ambient temperature until the preferred characteristics of *lait caillé* are obtained (Bayili et al., 2019; Savadogo et al., 2004). In Senegal, *lait caillé* is produced solely from cow's milk. The milk is first boiled (pasteurised) and then cooled before it is poured into a *lahal*, which is a wooden bowl that serves as the container for fermentation (Figure 4.2). The *lahals* are used in succession for fermentation, and over time they develop a biofilm layer due to the microbial culture from the previous fermentation cycle, which serves for passive back-slopping. The fermentation is conducted at 28°C for three days (Groenenboom et al., 2019).

In the Burkina Faso *lait caillé*, microbial successions indicated *Leuconostoc mesenteroides* to be the predominant LAB, followed by *Pediococcus pentosaceus* and *Weissella paramesenteroides* at the onset, while *Lactococcus lactis* and *Enterococcus* spp. were the predominant LAB after 7 hours of fermentation. The yeast *C. parapsilosis* dominated the first 18 hours of fermentation, while from 35 hours until the end of the fermentation *S. cerevisiae* dominated (Bayili et al., 2019). In the Senegalese *lait caillé*, the *Lactobacillus helveticus*, *Acetobacter pasteurianus* and *Lactococcus lactis* were the microflora of the *lahal*. Likewise, in the fermented *lait caillé*, the dominant LAB has been determined to be *Lactobacillus helveticus*, followed by *Acetobacter pasteurianus* and *Lactococcus lactis* (Groenenboom et al., 2019).

4.5.3 Warankasi (Wara)

Warankasi, commonly referred to as *wara*, is a traditional West African soft cheese made primarily from fresh cow's milk. This cheese is widely consumed in Nigeria, Ghana and Benin. Its production involves a simple yet microbiologically rich process, often using natural coagulants and traditional methods that contribute to its unique flavour and texture. The traditional production of *wara* involves collecting fresh cow's milk in calabashes, which is then transferred into a metal pot; the milk is heated at 45–50°C for 30 minutes, after which the coagulant is added. Coagulation is typically achieved using plant-based materials such as the extract from Sodom apple leaves (*Calotropis procera*) or leaves from the pawpaw (*Carica papaya*) plant (Adeyeye, 2017). Occasionally, lemon juice or vinegar is also used as a coagulant (Adetunji et al., 2008; Ogunlade et al., 2019). The microbiology of *wara* is diverse, containing various microorganisms that contribute to the cheese's characteristics. The main LAB in wara are the *Lactobacillus* and *Pediococcus* species; in particular *Lp. plantarum* and *Lm. fermentum* are predominant, followed by *P. pentosaceus* and *P. acidilactici* (Banwo et al., 2012).

In summary, a range of traditional fermented dairy products from various parts of Africa are displayed in the visual representation (Figure 4.3). The distinct fermenting microorganisms give each product its own flavour, texture and nutritional makeup.

4.6 Conclusion

African fermented dairy products, such as *amasi* from South Africa, *nunu* from Ghana, *ergo* from Egypt and *suusac* from Kenya, are rich in nutritional value and host a diverse microbiota that play a crucial role in their fermentation process. These products are often crafted using traditional methods which rely on natural (spontaneous) fermentation, contributing to the development of distinct sensory characteristics, including tangy flavours and creamy textures, giving them unique flavours, textures and health benefits. The microbiological landscape of these fermented dairy products is predominantly composed of LAB such as *Lactobacillus*, *Lactococcus*, *Streptococcus* and *Leuconostoc* species. These bacteria play a critical role in lowering the pH and preserving the dairy products, thereby inhibiting the growth of spoilage and pathogenic microorganisms. In addition to LAB, other beneficial microorganisms such as yeasts and moulds are occasionally involved.

The traditional methods of producing these fermented dairy products involve natural fermentation using raw milk and/or indigenous starter cultures, often derived from previous batches of fermented

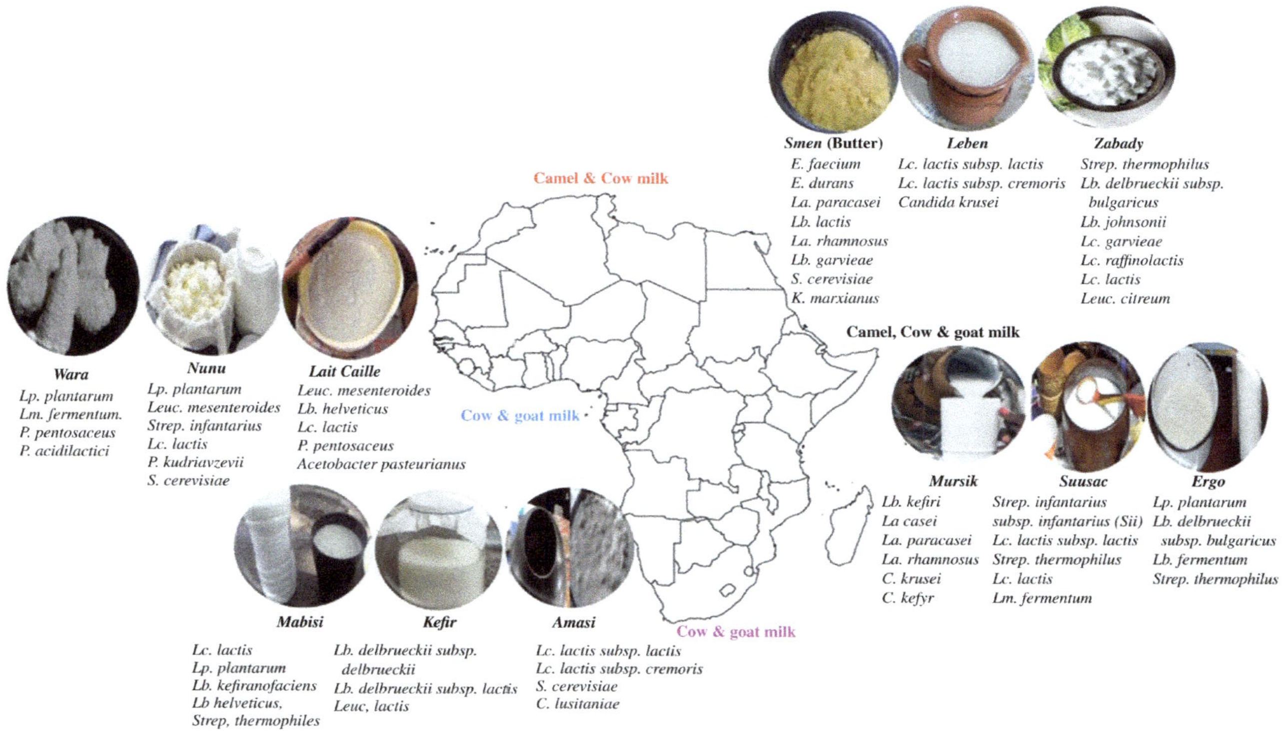

Figure 4.3 A pictorial view of various African fermented dairy products from different regions of Africa and the distinct fermenting microorganisms. (*Smen* (https://tinyurl.com/kydpjnft), *leben* (https://tinyurl.com/39jnfxva), *zabady* (https://tinyurl.com/39urdnta), *mursik* (https://tinyurl.com/3xekj9sa), *suusac* (From Kanetkar et al., 2023), *ergo* (From Kuyu & Bereka, 2020), *amasi* (From Maleke et al., 2021), *kefir* (https://tinyurl.com/4d8w5z84), *mabisi* (From Moonga et al., 2022), *lait caillé* (From Bayili et al., 2023) and *wara* (https://tinyurl.com/bddrs57n).)

milk. This method preserves the microbial diversity and heritage of the product, although it can result in variability in quality and safety. Fermented African dairy products are generally considered safe and beneficial due to the presence of LAB; however, the potential for contamination by pathogenic microorganisms exists. Ensuring the safety of these traditional foods requires a combination of good hygienic practices, proper fermentation techniques and regular monitoring to prevent foodborne illnesses and protect public health.

While the microbiota of many fermented African dairy-based products has been documented based on phenotypic techniques, further characterization using advanced molecular and metagenomic approaches is necessary to gain deeper insights into their microbial composition. This can assist with process control, rapid identification of potential pathogenic strains and the prevention of disease outbreaks associated with these products.

African fermented dairy products are a testament to the rich microbial biodiversity and traditional knowledge inherent in African cultures. They offer significant nutritional and health benefits due to the presence of beneficial microbes. However, there is a need to balance traditional practices with modern food safety and quality control measures to ensure these products can be enjoyed by broader populations without compromising their unique characteristics. This balance is essential for preserving the cultural heritage of these products while ensuring the health and safety of consumers.

References

Abdelgadir, W., Nielsen, D. S., Hamad, S., & Jakobsen, M. (2008). A traditional Sudanese fermented camel's milk product, Gariss, as a habitat of *Streptococcus infantarius* subsp. infantarius. *International Journal of Food Microbiology*, *127*(3), 215–219.

Abdelgadir, W. S., Ahmed, T. K., & Dirar, H. A. (1998). The traditional fermented milk products of the Sudan. *International Journal of Food Microbiology*, *44*(1–2), 1–13.

Abdelgadir, W. S., Hamad, S. H., Møller, P. L., & Jakobsen, M. (2001). Characterisation of the dominant microbiota of Sudanese fermented milk Rob. *International Dairy Journal*, *11*(1–2), 63–70.

Abou-Donia, S. A. (2004). Recent developments in Zabady and Egyptian Labneh research: A review. *Egyptian Journal of Dairy Science*, *32*, 1–15.

Adetunji, V. O., Alonge, D. O., Singh, R. K., & Chen, J. (2008). Production of wara, a West African soft cheese using lemon juice as a coagulant. *LWT – Food Science and Technology*, *41*(2), 331–336.

Adeyeye, S. A. O. (2017). A preliminary study on the quality and safety of street-vended warankasi (a Nigerian soft white cheese) from Ibadan, Oyo state, Nigeria. *British Food Journal*, *119*(2), 322–330.

Agyei, D., Owusu-Kwarteng, J., Akabanda, F., & Akomea-Frempong, S. (2020). Indigenous African fermented dairy products: Processing technology, microbiology and health benefits. *Critical Reviews in Food Science and Nutrition*, *60*(6), 991–1006.

Ajibola, W. G., Adeyemo, S. M., & Omafuvbe, B. O. (2020). Effect of starter culture fermentation of milk on the production, sensory attributes and storage of Wara (a Nigerian unripened soft cheese). *Ife Journal of Science*, *22*(3), 107–117.

Akabanda, F., Owusu-Kwarteng, J., Glover, R., & Tano-Debrah, K. (2010). Microbiological characteristics of Ghanaian traditional fermented milk product, *nunu*. *Nature and Science*, *8*, 178–187.

Akabanda, F., Owusu-Kwarteng, J., Tano-Debrah, K., Glover, R. L., Nielsen, D. S., & Jespersen, L. (2013). Taxonomic and molecular characterization of lactic acid bacteria and yeasts in *nunu*, a Ghanaian fermented milk product. *Food Microbiology*, *34*, 277–283.

Alnakip, M. E., El-Essely, E., Bayoumi, M. A., & Elsaid, E. I. (2023). Assigning the probiotic potential of lactic acid bacteria recovered from popular Egyptian fermented artisanal dairy products. *Journal of Advanced Veterinary Research*, *13*(7), 1443–1452.

Ashenafi, M. (1989). Proteolytic, lipolytic and fermentative properties of yeasts isolated from ayib, a traditional Ethiopian collage cheese. *SINET: Ethiopian Journal of Science*, *12*, 131–139.

Ashenafi, M. (1994). The aerobic microflora and lactic acid bacteria of market Ayib. *Ethiopian Journal of Agricultural Sciences*, *14*, 104–111.

Ashenafi, M. (1995). Microbial development and some chemical changes during the making of 'ergo'. A traditional Ethiopian sour milk. *Bulletin of Animal Health and Production in Africa*, *43*, 171–176.

Assefa, E., Beyene, F., & Santhanam, A. (2008). Isolation and characterization of inhibitory substance producing lactic acid bacteria from Ergo, Ethiopian traditional fermented milk. *Livestock Research for Rural Development*, *20*(3), 1–9.

Bamgbose, T. O., Sinha, S., Abdullahi, I. O., Inabo, H. I., Bello, M., Kori, L. D., Ametefe, E. N., & Anvikar, A. R. (2022). Identification of predominant lactic acid bacteria associated with kunun-zaki and kindirmo a traditional fermented food of Nigeria. *Current Topics in Lactic Acid Bacteria and Probiotics*, *8*(1), 17–31. https://doi.org/10.21203/rs.3.rs-1577271/v1

Banwo, K., Sanni, A., Tan, H., & Tian, Y. (2012). Phenotypic and genotypic characterization of lactic acid bacteria isolated from some Nigerian traditional fermented foods. *Food Biotechnology*, *26*(2), 124–142.

Bayili, G. R., Johansen, P., Nielsen, D. S., Sawadogo-Lingani, H., Ouedraogo, G. A., Diawara, B., & Jespersen, L. (2019). Identification of the predominant microbiota during production of *lait caillé*, a spontaneously fermented milk product made in Burkina Faso. *World Journal of Microbiology and Biotechnology*, *35*(100), 1–13.

Bayili, G. R., Konkobo-Yameogo, C., Diarra, S., Diawara, B., Jespersen, L., & Sawadogo-Lingani, H. (2023). Effect of fermentation process on hygiene and perceived quality of *lait caillé*, an ethnic milk product from Burkina Faso. *Journal of Ethnic Foods*, *10*(17), 1–13.

Bell, V., Guina, J., & Fernandes, T. H. (2023). African fermented foods and beverages. Potential impact on health. In *Microbial fermentations in nature and as designed processes* (pp. 293–322). John Wiley & Sons Ltd.

Bengoa, A. A., Iraporda, C., Garrote, G. L., & Abraham, A. G. (2019). Kefir micro-organisms: Their role in grain assembly and health properties of fermented milk. *Journal of Applied Microbiology*, *126*(3), 686–700.

Benkerroum, N. (2013). Traditional fermented foods of North African countries: Technology and food safety challenges with regard to microbiological risks. *Comprehensive Reviews in Food Science and Food Safety*, *12*(1), 54–89.

Benkerroum, N., & Tamime, A. Y. (2004). Technology transfer of some Moroccan traditional dairy products (lben, jben and smen) to small industrial scale. *Food Microbiology*, *21*(4), 399–413.

Bensalah, F., Delorme, C., & Renault, P. (2009). Characterisation of thermotolerant cocci from indigenous flora of 'leben' in Algerian arid area and DNA identification of atypical *Lactococcus lactis* strains. *Current Microbiology*, *59*(2), 139–146.

Berhe, T., Ipsen, R., Seifu, E., Kurtu, M. Y., Fugl, A., & Hansen, E. B. (2019). Metagenomic analysis of bacterial community composition in Dhanaan: Ethiopian traditional fermented camel milk. *FEMS Microbiology Letters*, *366*(11), fnz128.

Beukes, E. M., Bester, B. H., & Mostert, J. F. (2001). The microbiology of South African traditional fermented milks. *International Journal of Food Microbiology*, *63*(3), 189–197.

Boubekri, K., & Ohta, Y. (1996). Identification of lactic acid bacteria from Algerian traditional cheese, El-Klila. *Journal of the Science of Food and Agriculture*, *70*(4), 501–505.

Boussekine, R., Bekhouche, F., Debaets, S., Thierry, A., Maillard, M. B., Falentin, H., Pawtowski, A., Barkat, M., Coton, M., & Mounier, J. (2022). Deciphering the microbiota and volatile profiles of Algerian *smen*, a traditional fermented butter. *Microorganisms*, *10*(4), 736.

Boussekine, R., Merabti, R., Barkat, M., Becila, F.-Z., Belhoula, N., Mounier, J., & Bekhouche, F. (2020). Traditional fermented butter smen/dhan: Current knowledge, production and consumption in Algeria. *Journal of Food Research*, *9*, 71–82.

Digo, C. A., Kamau-Mbuthia, E., Matofari, J. W., & Ng'etich, W. K. (2017). Potential probiotics from traditional fermented milk, Mursik of Kenya. *International Journal of Nutrition and Metabolism*, *10*(9), 75–81.

Du Plooy, Z., Schönfeldt, H. C., & Hall, N. (2018). The role of traditional foods in food-based dietary guidelines – A South African case study on maas (cultured milk). *Food Chemistry*, *238*, 22–28.

El-Baradei, G., Delacroix-Buchet, A., & Ogier, J. C. (2008). Bacterial biodiversity of traditional Zabady fermented milk. *International Journal of Food Microbiology*, *121*(3), 295–301.

El-Gendy, S. (1983). Fermented foods of Egypt and the Middle East. *Journal of Food Protection*, *46*, 358–367.

Farah, Z., Mollet, M., Younan, M., & Dahir, R. (2007). Camel dairy in Somalia: Limiting factors and development potential. *Livestock Science*, *110*, 187–191.

Feresu, S. B., & Muzondo, M. I. (1990). Identification of some lactic acid bacteria from two Zimbabwean fermented milk products. *World Journal of Microbiology and Biotechnology*, *6*, 178–186.

Fox, P. F. (2009). Lactose: Chemistry and properties. In *Advanced dairy chemistry: Volume 3: Lactose, water, salts and minor constituents* (pp. 1–15). Springer.

Fugl, A., Berhe, T., Kiran, A., Hussain, S., Laursen, M. F., Bahl, M. I., Hailu, Y., Sørensen, K. I., Guya, M. E., Ipsen, R., & Hansen, E. B. (2017). Characterisation of lactic acid bacteria in spontaneously fermented camel milk and selection of strains for fermentation of camel milk. *International Dairy Journal*, *73*, 19–24.

Gadaga, T. H., Mutukumira, A. N., & Narvhus, J. A. (2000). Enumeration and identification of yeasts isolated from Zimbabwean traditional fermented milk. *International Dairy Journal*, *10*(7), 459–466.

Gebreselassie, N., Abay, F., & Beyene, F. (2016). Biochemical and molecular identification and characterization of lactic acid bacteria and yeasts isolated from Ethiopian naturally fermented buttermilk. *Journal of Food Science and Technology*, *53*, 184–196.

Ghosh, T., Beniwal, A., Semwal, A., & Navani, N. K. (2019). Mechanistic insights into probiotic properties of lactic acid bacteria associated with ethnic fermented dairy products. *Frontiers in Microbiology*, *10*, 502.

Gonfa, A., Fite, A., Urga, K., & Gashe, B. A. (1999). Microbiological aspects of Ergo (Ititu) fermentation. *SINET: Ethiopian Journal of Science*, *22*(2), 283–290.

Gonfa, A., Foster, H. A., & Holzapfel, W. H. (2001). Field survey and literature review on traditional fermented milk products of Ethiopia. *International Journal of Food Microbiology*, *68*(3), 173–186.

Groenenboom, A. E., Parker, M. E., De Vries, A., De Groot, S., Zobrist, S., Mansen, K., Milani, P., Kort, R., Smid, E. J., & Schoustra, S. E. (2019). Bacterial community dynamics in lait caillé, a traditional product of spontaneous fermentation from Senegal. *PLoS One*, *14*(5), e0215658.

Groenenboom, A. E., Shindano, J., Cheepa, N., Smid, E. J., & Schoustra, S. E. (2020). Microbial population dynamics during traditional production of Mabisi, a spontaneous fermented milk product from Zambia: A field trial. *World Journal of Microbiology and Biotechnology*, *36*, 1–14.

Guetouache, M., & Guessas, B. (2015). Characterization and identification of lactic acid bacteria isolated from traditional cheese (Klila) prepared from cows milk. *African Journal of Microbiology Research*, *9*(2), 71–77.

Hawaz, H., Bottari, B., Scazzina, F., & Carini, E. (2025). Eastern African traditional fermented foods and beverages: Advancements, challenges, and perspectives on food technology, nutrition, and safety. *Comprehensive Reviews in Food Science and Food Safety*, *24*(2), e70137.

Igwe, E. C., Esiobu, E. N., & Ojimelukwe, P. C. (2014). Variations in the traditional starter culture for production of a Nigerian fermented milk product (kindirmo). *Focusing on Modern Food Industry*, *3*, 35.

Iradukunda, C., Aida, W. M. W., Ouafi, A. T., Barkouch, Y., & Boussaid, A. (2018). Aroma profile of a traditionally fermented butter (smen). *Journal of Dairy Research*, *85*, 114–120.

Jans, C., Bugnard, J., Njage, P. M. K., Lacroix, C., & Meile, L. (2012). Lactic acid bacteria diversity of African raw and fermented camel milk products reveals a highly competitive, potentially health-threatening predominant microflora. *LWT*, *47*(2), 371–379.

Jespersen, L. (2003). Occurrence and taxonomic characteristics of strains of Saccharomyces cerevisiae predominant in African indigenous fermented foods and beverages. *FEMS Yeast Research*, *3*(2), 191–200.

Jiwoua, C., & Milliére, J. B. (1990). Lactic flora and enterococci of fermented [zebu] milk (pindidam) produced in Adamaoua (Cameroun). *Lait (France)*, *70*(5–6), 475–486.

Kanetkar, P., Paswan, V. K., Rose, H., Shehata, A. M., Felix, J., Bunkar, D. S., Rathaur, A., Yamini, S., & Bhinchhar, B. K. (2023). Appraisal of some ethnic milk products from minor milch animal species around the world: A review. *Journal of Ethnic Foods*, *10*(1), 40.

Karenzi, E., Mashaku, A., Nshimiyimana, A. M., Munyanganizi, B. T., & Thonart, P. (2013). Kivuguto traditional fermented milk and the dairy industry in Rwanda. A review. *Biotechnologie, agronomie, société et environnement*, *17*(2), 383–391.

Kirui, E., & Nguka, G. (2017). Factors that negatively influence consumption of traditionally fermented milk (Mursik) among preschool children (1–5 years old) in Kapseret location-Uasin Gishu County, Kenya. *African Journal of Food, Agriculture, Nutrition and Development*, *17*(3), 12295–12310.

Kuyu, C. G., & Bereka, T. Y. (2020). Review on contribution of indigenous food preparation and preservation techniques to attainment of food security in Ethiopian. *Food science & nutrition*, *8*(1), 3–15.

Labtar, A., Larouci, S., Guermouche, A., & Bensalah, F. (2019). Study on molecular identification of lactic acid bacteria from fermented milks and “Smen” (a traditional steppe butter) and their enzyme producing attributes. *South Asian Journal of Experimental Biology*, *9*(3).

Lelise, A., Belaynesh, G., Mulubrhan, M., Kedija, S., Endashaw, B., & Abebe, B. (2014). Isolation and screening of antibacterial producing lactic acid bacteria from traditionally fermented drinks (“Ergo” and

"Tej") in Gondar town, Northwest Ethiopia. *Global Research Journal of Public Health and Epidemiology*, *1*(3), 18–22.

Lore, T. A., Mbugua, S. K., & Wangoh, J. (2005). Enumeration and identification of microflora in suusac, a Kenyan traditional fermented camel milk product. *LWT – Food Science and Technology*, *38*(2), 125–130.

Loretan, T., Mostert, J. F., & Viljoen, B. C. (2003). Microbial flora associated with South African household kefir. *South African Journal of Science*, *99*(1), 92–94.

Maleke, M. S., Adefisoye, M. A., Doorsamy, W., & Adebo, O. A. (2021). Processing, nutritional composition and microbiology of amasi: A Southern African fermented milk product. *Scientific African*, *12*, e00795.

Mathara, J. M., Schillinger, U., Kutima, P. M., Mbugua, S. K., & Holzapfel, W. H. (2004). Isolation, identification and characterisation of the dominant microorganisms of kule naoto: The Maasai traditional fermented milk in Kenya. *International Journal of Food Microbiology*, *94*(3), 269–278.

Mattiello, S., Caroprese, M., Matteo, C. G., Fortina, R., Martini, A., Martini, M., Parisi, G., Russo, C., Zecchini, M., & ASPA Commission "Animal Productions in Development Cooperation Projects". (2018). Typical dairy products in Africa from local animal resources. *Italian Journal of Animal Science*, *17*(3), 740–754.

Mbawala, A., Mahbou, P. Y., Mouafo, H. T., & Tatsadjieu, L. N. (2013). Antibacterial activity of some lactic acid bacteria isolated from a local fermented milk product (pendidam) in ngaoundere, Cameroon. *Journal of Animal and Plant Sciences*, *23*(1), 157–166.

Moonga, H. B., Schoustra, S. E., Linnemann, A. R., Kuntashula, E., Shindano, J., & Smid, E. J. (2019). The art of mabisi production: A traditional fermented milk. *PLoS ONE*, *14*(3), e0213541.

Moonga, H. B., Schoustra, S. E., Linnemann, A. R., Shindano, J., & Smid, E. J. (2022). Towards valorisation of indigenous traditional fermented milk: Mabisi as a model. *Current Opinion in Food Science*, *46*, 100835.

Moonga, H. B., Schoustra, S. E., Van den Heuvel, J., Linnemann, A. R., Samad, M. S., Shindano, J., & Smid, E. J. (2020). Composition and diversity of natural bacterial communities in mabisi, a traditionally fermented milk. *Frontiers in Microbiology*, *11*, 1816.

Mourad, G., & Bettache, G. (2019). A brief summary scheme of Algerian traditional dairy products. *Journal of Food Quality and Hazards Control*, *6*(3), 80–81.

Nahaisi, M., Almaroum, N., & Ziyaina, M. (2024). Isolation and characterization of lactic acid bacteria from Libyan traditional fermented Milk "laban". *The Libyan Journal of Agriculture*, *28*(2).

Nakavuma, J. L., Møller, P. L., Jakobsen, M., Salimo, P., & Nasinyama, G. W. (2012). Processing steps and lactic acid bacteria involved in traditional cultured milk (Kwerionik) production in Uganda. *Africa Journal of Animal and Biomedical Sciences*, 7(2), 82–93.

Narvhus, J. A., & Gadaga, T. H. (2003). The role of interaction between yeasts and lactic acid bacteria in African fermented milks: A review. *International Journal of Food Microbiology*, *86*(1–2), 51–60.

Nieminen, M. T., Novak-Frazer, L., Collins, R., Dawsey, S. P., Dawsey, S. M., Abnet, C. C., White, R. E., Freedman, N. D., Mwachiro, M., Bowyer, P., Salaspuro, M., & Rautemaa, R. (2013). Alcohol and acetaldehyde in African fermented milk mursik—A possible etiologic factor for high incidence of esophageal cancer in western Kenya. *Cancer Epidemiology, Biomarkers & Prevention*, *22*(1), 69–75.

Nyambane, B., Thari, W. M., Wangoh, J., & Njage, P. M. (2014). Lactic acid bacteria and yeasts involved in the fermentation of amabere amaruranu, a Kenyan fermented milk. *Food Science & Nutrition*, *2*(6), 692–699.

Obodai, M., & Dodd, C. E. R. (2006). Characterization of dominant microbiota of a Ghanaian fermented milk product, nyarmie, by culture- and nonculture-based methods. *Journal of Applied Microbiology*, *100*(6), 1355–1363.

Ogunlade, A. O., Oyetayo, V. O., Ojokoh, A. O., & Aladejana, O. M. (2019). Influence of different coagulants on the amino acid content of soft cheese (wara) produced from sheep milk. *Quest Journals Journal of Research in Agriculture and Animal Science*, *6*(1), 35–39.

Ongol, M. P., & Asano, K. (2009). Main microorganisms involved in the fermentation of Ugandan ghee. *International Journal of Food Microbiology*, *133*(3), 286–291.

Ouadghiri, M., Vancanneyt, M., Vandamme, P., Naser, S., Gevers, D., Lefebvre, K., Swings, J., & Amar, M. (2009). Identification of lactic acid bacteria in Moroccan raw milk and traditionally fermented skimmed milk 'lben'. *Journal of Applied Microbiology*, *106*(2):486–495.

Owusu-Kwarteng, J., Akabanda, F., Agyei, D., & Jespersen, L. (2020). Microbial safety of milk production and fermented dairy products in Africa. *Microorganisms*, *8*(5), 752.

Owusu-Kwarteng, J., Akabanda, F., Johansen, P., Jespersen, L., & Nielsen, D. S. (2017). Nunu, a West African fermented yogurt-like milk product. In *Yogurt in health and disease prevention* (pp. 275–283). Academic Press.

Plessas, S., Nouska, C., Mantzourani, I., Kourkoutas, Y., Alexopoulos, A., & Bezirtzoglou, E. (2016). Microbiological exploration of different types of kefir grains. *Fermentation*, *3*(1), 1.

Saleh, F. A. (2013). Isolation and identification of microorganisms and antibacterial activity of Laban Zeer, an Egyptian traditional fermented milk product. *Scientific Journal of Microbiology*, *2*(2), 31–42.

Samet-Bali, O., Ayadi, M. A., & Attia, H. (2012). Development of fermented milk "Laban" made from spontaneous fermented cow's milk. *African Journal of Biotechnology*, *11*(7), 1829–1837.

Samet-Bali, O., Felfoul, I., Lajnaf, R., Attia, H., & Ayadi, M. (2016). Hygienic quality of "Rayeb", a traditional Tunisian fermented cow's milk. *International Food Research Journal*, *23*(1): 366–369.

Sangoyomi, T. E., Owoseni, A. A., & Okerokun, O. (2010). Prevalence of enteropathogenic and lactic acid bacteria species in wara: A local cheese from Nigeria. *African Journal of Microbiological Reserach*, *4*(15), 1624–1630.

Sarhir, S. T., Amanpour, A., Bouseta, A., & Selli, S. (2019). Key odorants of a Moroccan fermented milk product "Lben" using aroma extract dilution analysis. *Journal of Food Science and Technology*, *56*(8), 3836–3845.

Savadogo, A., Ouattara, C. A. T., Savadogo, P. W., Ouattara, A. S., Barro, N., & Traore, A. S. (2004). Microorganisms involved in Fulani traditional fermented milk in Burkina Faso. *Pakistan Journal of Nutrition*, *3*(2), 134–139.

Schoustra, S. E., Kasase, C., Toarta, C., Kassen, R., & Poulain, A. J. (2013). Microbial community structure of three traditional Zambian fermented products: Mabisi, chibwantu and munkoyo. *PLoS ONE*, *8*(5), e63948.

Schutte, L. M. (2013). *Isolation and identification of the microbial consortium present in fermented milks from sub-Saharan Africa* [Doctoral dissertation, Stellenbosch: Stellenbosch University.]

Shori, A. B. (2012). Comparative study of chemical composition, isolation and identification of micro-flora in traditional fermented camel milk products: Gariss, Suusac, and Shubat. *Journal of the Saudi Society of Agricultural Sciences*, *11*(2), 79–88.

Simatende, P., Siwela, M., & Gadaga, T. H. (2019). Identification of lactic acid bacteria and determination of selected biochemical properties in emasi and emahewu. *South African Journal of Science*, *115*(11–12), 1–7.

Sulieman, A. M. E. H., Ilayan, A. A., & Faki, A. E. A. F. (2006). Chemical and microbiological quality of Garris, Sudanese fermented camel's milk product. *International Journal of Food Science and Technology*, *41*(3), 321–328.

Tantaoui-Elaraki, A., & El Marrakchi, A. (1987). Study of Moroccan dairy products: Iben and smen. *MIRCEN Journal of Applied Microbiology and Biotechnology*, *3*, 211–220.

Todorov, S. D., Nyati, H., Meincken, M., & Dicks, L. M. T. (2007). Partial characterization of bacteriocin AMA-K, produced by *Lactobacillus plantarum* AMA-K isolated from naturally fermented milk from Zimbabwe. *Food Control*, *18*(6), 656–664.

Vorster, H. H., Wentzel-Viljoen, E., & Vermaak, M. (2013). "Have milk, maas or yoghurt every day": A food-based dietary guideline for South Africa. *South African Journal of Clinical Nutrition*, *26*, S57–S65.

Walsh, A. M., Crispie, F., Daari, K., O'Sullivan, O., Martin, J. C., Arthur, C. T., Claesson, M. J., Scott, K. P., & Cotter, P. D. (2017). Strain-level metagenomic analysis of the fermented dairy beverage nunu highlights potential food safety risks. *Applied and Environmental Microbiology*, *83*(16), e01144–17.

Witthuhn, R. C., Schoeman, T., & Britz, T. J. (2004). Isolation and characterization of the microbial population of different South African kefir grains. *International Journal of Dairy Technology*, *57*(1), 33–37.

Wullschleger, S., Lacroix, C., Bonfoh, B., Sissoko-Thiam, A., Hugenschmidt, S., Romanens, E., Baumgartner, S., Traoré, I., Yaffee, M., Jans, C., & Meile, L. (2013). Analysis of lactic acid bacteria communities and their seasonal variations in a spontaneously fermented dairy product (Malian fènè) by applying a cultivation/genotype-based binary model, *International Dairy Journal*, *29*(1), 28–35.

5

Microbiology of African Fermented Roots and Tubers

5.1 Introduction

In Africa, fermented root and tuber products are an integral part of traditional diets in many regions of the continent. These products primarily incorporate fermented cassava (*Manihot esculenta*), yams (*Dioscorea* spp.) and sweet potatoes (*Ipomoea batatas*), which are staple foods in various African communities. These dishes require preparing roots and tubers using traditional fermentation processes involving diverse microbial communities, which play a crucial role in enhancing their safety, nutritional value and sensory properties (Oyewole & Odunfa, 1992; Padonou, Hounhouigan, et al., 2009).

Fermented cassava products are known by various local names such as *gari*, *agbelima*, *attieke*, *fufu* and *kokonte* in West Africa, while Central Africa produces products such as *chikwangue*, *kwanga* and *kivunde* (Akintayo et al., 2023). The diversity of African fermented root and tuber-based products is listed in Table 5.1. Cassava, one of the most widely fermented roots, is prepared using a process that significantly reduces its cyanogenic glucoside content, thus detoxifying the tuber (Panghal et al., 2019). Typically, cassava processing involves peeling, washing, soaking in water and fermenting cassava tubers submerged in water for several days to reduce the cyanogenic compounds (Akintayo et al., 2023; Padonou, Hounhouigan, et al., 2009). There are variations to the fermentation processes, which have been broadly categorised into a submerged fermentation process, which involves the soaking of the roots under water (retting), and a solid-state fermentation (SSF) process, which does not include soaking (Oyewole, 1995). In regions where yams are a staple, fermentation reduces postharvest losses and extends shelf

DOI: 10.1201/9781003658320-5

Table 5.1 A List of African Fermented Root and Tuber-Based Products

PRODUCT	ROOT AND TUBER USED	PRODUCT FORM	COUNTRY/REGION
Abacha	Cassava	Dish	Nigeria
Agbelima	Cassava	Meal	Ivory Coast and Ghana
Attiéké/Placali	Cassava	Meal	Ivory Coast
Bikedi	Cassava	Meal	DR Congo
Chikwangue/Kwanga	Cassava	Meal, dish	Central Africa, Zaire
Cingwada	Cassava	Meal, dish	East and Central Africa
Cossettes	Cassava	Flour	DR Congo and Rwanda
Elubo isu/Gbodo	Yam	Flour	Nigeria
Fufu	Cassava	Flour	West Africa
Gari	Cassava	Flour	West and Central Africa
Inyanga	Cassava	Flour	Burundi
Ikivunde	Cassava	Flour	Burundi
Kanyanga	Cassava	Distilled spirit	Burundi
Kivunde	Cassava	Meal	Tanzania
Kokobele	Cocoyam corms	Meal	Nigeria
Konkonte	Cassava	Flour	Ghana
Kpor umilin	Cassava	Flour	Nigeria
Lafun	Cassava	Flour	Nigeria, Benin
Meduame-M-bong	Cassava	Meal	Cameroon
Mokopa	Cassava	Flour	Uganda, Tanzania, Rwanda, DR Congo
Mouyondo/Mpita	Cassava	Cassava sticks	Gabon
Placali	Cassava	Meal	Ivory Coast
Pupuru	Cassava	Meal	Nigeria
Usi	Cassava	Meal, Edible starch	Nigeria

Source: Adapted from Akintayo et al. (2023).

life (Ferraro et al., 2016). This method not only preserves the yams but also breaks down certain complex carbohydrates, making them easier to digest. Fermented yam products such as *amala* are staples in various African cuisines. Fermentation of sweet potatoes is less common but is practised in certain areas (Ajayi et al., 2016). The process typically involves slicing the sweet potatoes and fermenting them in a controlled environment to develop a distinctive sour flavour and improve digestibility. The fermentation process in root and tuber products is primarily driven by lactic acid bacteria (LAB), such as *Lactobacillus* spp. and *Leuconostoc* spp., along with yeast species like *Saccharomyces cerevisiae*, which play a crucial role in enhancing the palatability and nutritional profile by increasing the bioavailability of nutrients (Panda & Ray, 2016). These microorganisms are responsible for producing organic acids, which lower the pH and inhibit spoilage microorganisms, thereby enhancing the safety and shelf life of the product. The fermentation also improves the palatability and nutritional profile by increasing the bioavailability of essential nutrients. These fermented products are important for food security and contribute to the unique culinary traditions and flavours of African cuisine (Ajayi et al., 2016; Akintayo et al., 2023; Panda & Ray, 2016). They are a testament to the ingenuity of traditional food processing techniques passed down through generations.

The interactions between microbial species during the fermentation of roots and tubers are complex and synergistic. LAB dominate the initial stages of fermentation, producing organic acids that lower the pH. This acidic environment favours the growth of yeasts and other acid-tolerant microorganisms (Padonou, Nielsen, et al., 2009; Oyewole & Odunfa, 1992). These microbes collectively contribute to the flavour, texture and safety of the fermented products. The consumption of fermented root and tuber products has been associated with various health benefits. The reduced levels of cyanogenic compounds in fermented cassava minimise the risk of cyanide poisoning, a significant concern in regions where cassava is a dietary staple. Fermentation of root and tuber products plays a vital role in enhancing their overall safety, nutritional value and sensory attributes. Therefore, it is essential to elucidate the microbiology of these products to have a standardised process control during their production.

5.2 Selected African Fermented Root and Tuber-Based Products

5.2.1 Gari

Gari is a fermented product produced in West Africa by the spontaneous fermentation of cassava through SSF. *Gari* is a ready-to-eat, coarse dry granular flour that can be easily consumed by rehydrating it with water or mixing it into a jelly paste with hot water. The production process involves peeling fresh cassava tubers, grating the peeled tubers into a mash and then fermenting this mash in a jute bag for three to five days (Figure 5.1). After fermentation, the mash

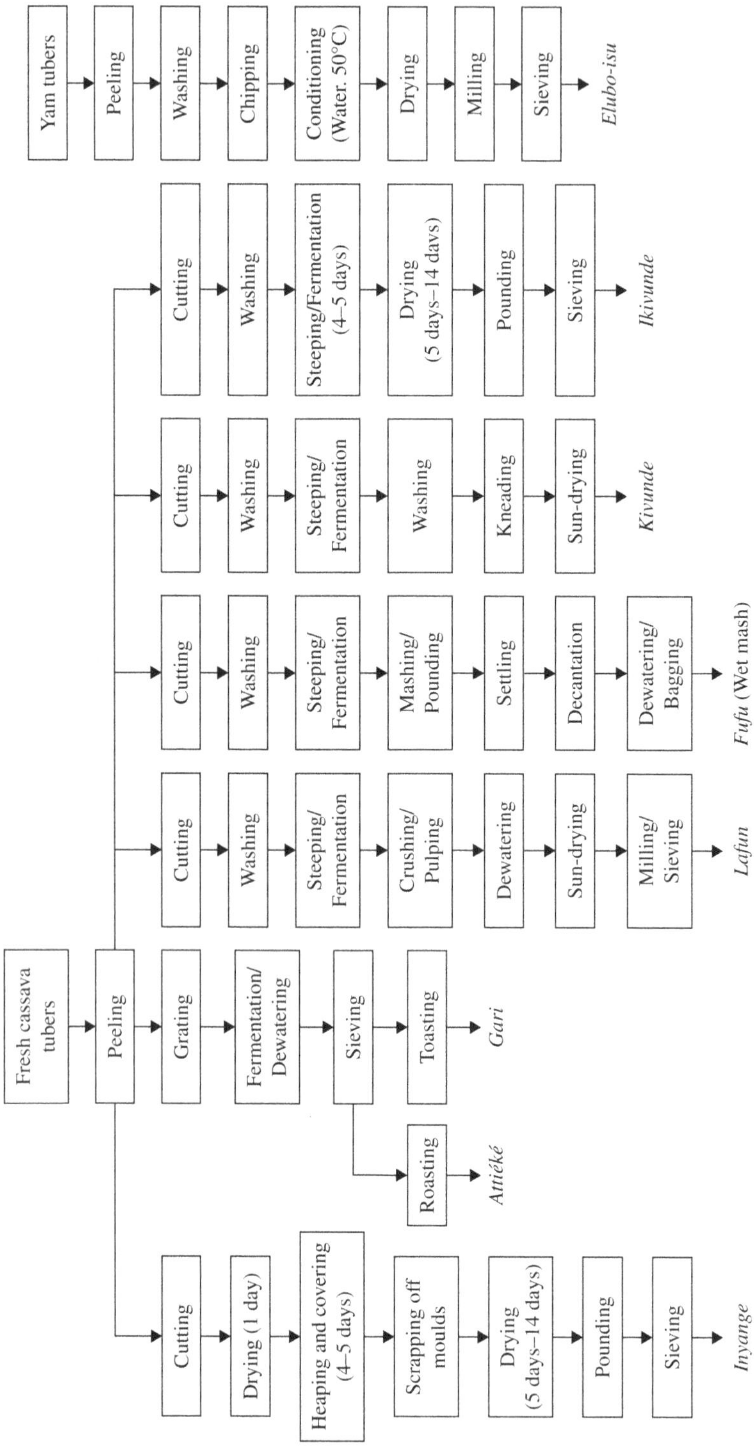

Figure 5.1 Schematic flow chart of the production of various African fermented products from cassava and yam.

is dehydrated by pressing, sieving and toasting it into a dry granular flour (Oguntoyinbo, 2007).

The predominant LAB in fermenting cassava mash is *Lactiplantibacillus plantarum*, followed by *Leuconostoc fallax* and *Limosilactobacillus fermentum* (Kostinek et al., 2005; Oguntoyinbo, 2007). Other LAB strains that appear occasionally are *Levilactobacillus brevis*, *Leuconostoc pseudomesenteroides* and *Alcaligenes faecalis*, *Leuconostoc cremoris* and *Geotrichum candidum*. The *Lp. plantarum* is solely responsible for producing lactic acid, which lowers the pH and inhibits the growth of spoilage organisms and pathogens. The ability of LAB to produce linamarase for detoxifying the cyanide present in cassava is essential in preparing cassava food products. The most predominant linamarase LAB produced by traditional cassava fermentation are the *Lactiplantibacillus plantarum*, *Lm. fermentum*, *Lv. brevis*, *Leuconostoc pseudomesenteroides*, *Lc. fallax* and *Weissella paramesenteroides* (Kostinek et al., 2005; Panda & Ray, 2016). Other essential properties of the fermenting LAB are their capability to produce antagonistic substances such as H_2O_2 and bacteriocins to inhibit the growth of pathogenic microorganisms and their ability to utilise complex sugars such as raffinose and stachyose during fermentation (Behera & Ray, 2017; Kostinek et al., 2005; Ogbonnaya, 2016). The diversity of microorganisms found in cassava fermentation to produce *gari* is listed in Table 5.2. Sanni et al. (2001) investigated the application of *Lactiplantibacillus plantarum* as a starter culture in cassava fermentation for *gari* production. The resulting *gari* had a lower pH and significantly higher lactic acid content. Fungi play an important role in *gari* fermentation. For instance, Akindahunsi and Oboh (2003) studied the impact of using a pure strain of *Rhizopus oryzae* in fermenting cassava pulp, which was later processed into *gari*. This process resulted in a *gari* with an improved protein content and lower levels of antinutrient factors. To further enhance the nutritional quality of *gari*, some researchers have employed *Saccharomyces cerevisiae* in the SSF of cassava pulp (Aguoru et al., 2014; Oboh & Akindahunsi, 2003) to produce a final product exhibiting reduced cyanide levels and increased protein and fat contents. Oguntoyinbo and Dodd (2010) examined the microbial dynamics and diversity during SSF of cassava for *gari* production in West Africa. Utilising advanced molecular techniques such as 16S rDNA gene sequence

Table 5.2 Major Fermenting Microorganisms of African Fermented Root and Tuber-Based Products

PRODUCT	MICROBE RESPONSIBLE FOR FERMENTATION	REFERENCES
Abacha	*Lactobacillus* spp., *Candida* spp., *Aspergillus* spp., *Bacillus* spp.	Ihenetu et al. (2017)
Agbelima	*Lactobacillus* spp. *(Levilactobacillus brevis and Lactiplantibacillus plantarum), Leuconostoc mesenteroides, Bacillus* spp. *(Bacillus subtilis and Bacilus licheniformis, Bacillus cereus, Bacillus mycoides and Bacillus polymixa), Penicillium* spp., *Candida tropicalis, Candida crusei, Zygosaccharomyces*	Amoa-Awua et al. (1996), Amoa-Awua et al. (1997)
Attiéké	*Leuconostoc mesenteroides* subsp. *mesenteroides, Ligilactobacillus salivarius, Lb. delbrueckii* subsp. *delbrueckii, Lp. plantarum, Lm. fermentum, Lb. confusus, Lv. brevis, Lactobacillus buchneri, Lactobacillus confusus, Lactobacillus delbrueckii, Leuc. mesenteroides, Weissella cibaria, W. confusa, Bacillus circulans, Bacillus lentus, C. tropicalis, Candida, krusei, Sacharomyces cerevisiae*	Coulin et al. (2006), Assanvo et al. (2006), Djeni et al. (2015), Djeni et al. (2015), Krabi et al. (2016)
Chikwangue/Kwanga	*Streptococcus faecalis and Bacillus licheniformis*	Ray and Ward (2006)
Elubo isu/ Amala	*B. subtilis, Candida krusei, Citrobacter freundii, Lv. brevis, Lactobacillus, Delbrueckii, Lp. plantarum, Lactococcus lactis, K.pneumoniae, Pichia burtonii*	Achi and Akubor (2000), Osho et al. (2010)
Fufu	*Lactiplantibacillus plantarum, Lactobacillus acidophilus, Lm. fermentum, Lb. cellobiosus, Lv. brevis, Lb. coprophilus, Lb. lactis, Lb. bulgaricus, Leuconostoc mesenteroides, Leuc. lactis, B. subtiflis, Bacillus polymyxa, Candida krusei, C. tropicalis, Pichia saitoi, P. anomala, Saccharomyces cerevisiae, Zygosaccharomyces bailii*	Adegoke and Babalola (1988), Oyewole and Odunfa (1990), Oyewole (2001), Oyedeji et al. (2013)
Gari	*Lactiplantibacillus plantarum, Limosilactobacillus fermentum, Levilactobacillus brevis, Leuconostoc fallax, Leuconostoc pseudomesenteroides, Weissella paramesenteroides*	Kostinek et al. (2005), Oguntoyinbo (2007)

(Continued)

Table 5.2 Major Fermenting Microorganisms of African Fermented Root and Tuber-Based Products *(Continued)*

PRODUCT	MICROBE RESPONSIBLE FOR FERMENTATION	REFERENCES
Iyange	*Aspergillus oryzae*, *Aspergillus fumigatus*, *Penicillium citrinum*, *Penicillium chrysogenum*, *Rhizopus stolonifer* and *Mucor* spp., *Geotrichum candidum*	Nzigamasabo and Hui Ming (2006a), Nzigamasabo and Hui Ming (2006b), Nzigamasabo and Nimpagaritse (2009)
Ikivunde	*Lactiplantibacillus plantarum*, *Levilactobacillus brevis*, *Limosilactobacillus fermentum*, *Leuconostoc mesenteroides*, *Geotrichum candidum*	Nzigamasabo and Hui Ming (2006a), Nzigamasabo and Hui Ming (2006b), Nzigamasabo and Nimpagaritse (2009)
Kivunde	*Lactoplantibacillus plantrum*, *Lactobacillus* spp.	Ray and Ward (2006), Kimaryo et al. (2000)
Kpor umilin	*Aspergillus* spp., *Bacillus* spp., *Enterobacter* spp. *E. coli*, *Geotrichum* spp., *Lactobacillus* spp., *Leuconostoc* spp., *Micrococcus* spp., *Penicillium* spp., *Saccharomyces* spp., *Staphylococcus* spp. *Streptococcus* spp.	Tsav-Wua et al. (2004)
Lafun	*Limosilactobacillus fermentum*, *Lactoplantibacillus plantarum*, *Weissella confusa*, *Saccharomyces cerevisiae*, *Pichia scutulata*, *Kluyveromyces marxianus*, *Hanseniaspora guilliermondii*, *Pichia rhodanensis*, *Candida glabrata*, *Pichia kudriavzevii*, *Candida tropicalis*, *Trichosporon asahii*, *Leuconostoc* spp., *Corynebacterium* spp., *Lactobacillus* spp., *Bacillus* spp., *Candida* spp., *Pichia* spp., *Rhodotorula* spp., *Hansenula* spp., *Geotrichum* spp., *Aspergillus* spp., *Penicillium* spp.	Padonou, Nielsen, et al. (2009)
Pupuru	*Alcaligenes* spp.; *Bacillus* sp., *Candida famata*; *Cryptococcushumicola*; *Geotrichum capitatum*; *Lactocaseibacillus casei*; *Lb. delbrueckii*; *Lv. brevis*; *Lp. plantarum*; *Lactobacillus pentosus*; *Lm. fermentum*; *Lb. acidophilus*; *Leuconostoc* spp.; *Serratia* spp.; *Saccharomyces* spp.	Wakil and Benjamin (2015), Adeyemo et al. (2018)

Abbreviations: La., Lacticaseibacillus; Lb., Lactobacillus; Li., Ligilactobacillus; Lm., Limosilactobacillus; Lp., Lactiplantibacillus; Lv., Levilactobacillus.

analysis and pulsed-field gel electrophoresis (PFGE), they identified several lactic acid bacterial species, including *Lactiplantibacillus plantarum*, *Limosilactobacillus fermentum*, *Lactobacillus pentosus*, *Lactobacillus acidophilus* and *Lactobacillus casei*. Ahaotu et al. (2013) discuss the microbial contributions to reducing toxicity and enhancing the nutritional profile of *gari* by fermentation of cassava pulp under water, highlighting the microbial linamarase activity from organisms such as *Alcaligenes faecalis*, *Lactiplantibacillus plantarum*, *Leuconostoc cremoris* and *Geotrichum candidum*, which are all associated with the breakdown of cyanogenic glucosides. The microbiology of *gari* is crucial to its production process, as the fermentation significantly affects its safety, flavour, texture and nutritional properties. The interplay between LAB, yeasts and other microorganisms during fermentation transforms cassava into a safe, nutritious and culturally important food product.

5.2.2 Fufu

Fufu is a traditional West African staple food generally made from cassava, although other variants can be made from cassava and/or plantain (Adu-Gyamfi, 2022; Oyewole & Odunfa, 1990). The primary ingredients for *fufu* vary depending on the country, with cassava being the most common. It is the second-most popular African fermented food after *gari*. *Fufu* is made by retting (soaking) the peeled cassava roots for three to five days (until the roots are softened), removing fibres and unsoftened parts of the roots and leaving the resulting mash in permeable sacks or strainers to allow the drainage of excess water, after which it is milled into a wet paste (*fufu*), then cooked over a fire (Figure 5.1). The preparation of *fufu* is a submerged fermentation process because the roots are soaked under water.

Bacillus species have been identified in the early stages of fermentation, with *Bacillus subtilis* and *Bacillus polymyxa* being the most prominent (Braide et al., 2018; Oyedeji et al., 2013). *Bacillus species* produce pectinase and other amylolytic enzymes to disintegrate the pectin in the cell wall of the cassava roots (Adesulu & Awojobi, 2014). Following pectinase activity, the LAB acts on the cassava roots to produce flavour. Oyedeji et al. (2013) found *Lactobacillus acidophilus*,

Levilactobacillus brevis, *Lactobacillus coprophillus*, *Lactobacillus lactis*, *Lactiplantibacillus plantarum* and *Leuconostoc mesenteroides* to be the predominant organisms during fermentation. Other studies have identified *Lactiplantibacillus plantarum* and *Limosilactobacillus fermentum* as the most dominant lactic acid bacterial strains involved in the fermentation of *fufu* (Babatuyi et al., 2020; Braide et al., 2018). Amylolytic yeasts such as *Candida krusei*, *C. tropicalis*, *Pichia saitoi*, *Saccharomyces cerevisiae*, *P. anomala* and *Zygosaccharomyces bailii* have been found in fermenting pulp (Oyewole, 2001). These yeasts produce amylolytic enzymes necessary for breaking down the complex sugars in cassava. Table 5.2 shows the assortment of fermenting microorganisms during *fufu* production.

The fermentation is primarily driven by the successional growth of LAB and yeasts. The LAB convert sugars in the cassava into lactic acid through anaerobic metabolism. This acidification lowers the pH of the mixture, creating a favourable environment for the subsequent action of other microorganisms. The yeasts, such as species of *Saccharomyces* and *Candida*, contribute to the development of flavours and aromas and other volatile compounds. While the population of LAB increases quantitatively as the fermentation progresses, a gradual reduction in the number of yeast groups present is to be expected. *Lactobacillus lactis* and *Lp. plantarum* have been found to dominate the fermentation after 60 hours, while the only yeast cultures identified at this point were *C. krusei* (Oyewole, 2001; Oyewole & Odunfa, 1990). The disappearance of yeasts could be attributed to the increasing acidity. The latter part of the fermentation is usually dominated by *Lp. plantarum*. The yeast *C. krusei* has been reported to significantly influence the production of the typical odour of *fufu* (Oyewole, 2001). The synergistic activity of these microbial communities transforms the raw cassava into *fufu*, a popular and culturally significant staple food in many African countries. The microbiota involved in *fufu* fermentation may vary depending on factors such as geographical location and traditional fermentation practices. Overall, the microbial activities during *fufu* production are pivotal in achieving the desired sensory attributes, shelf stability and nutritional quality of the final product. Figure 5.2 shows a pictorial view of various African fermented roots and tuber products.

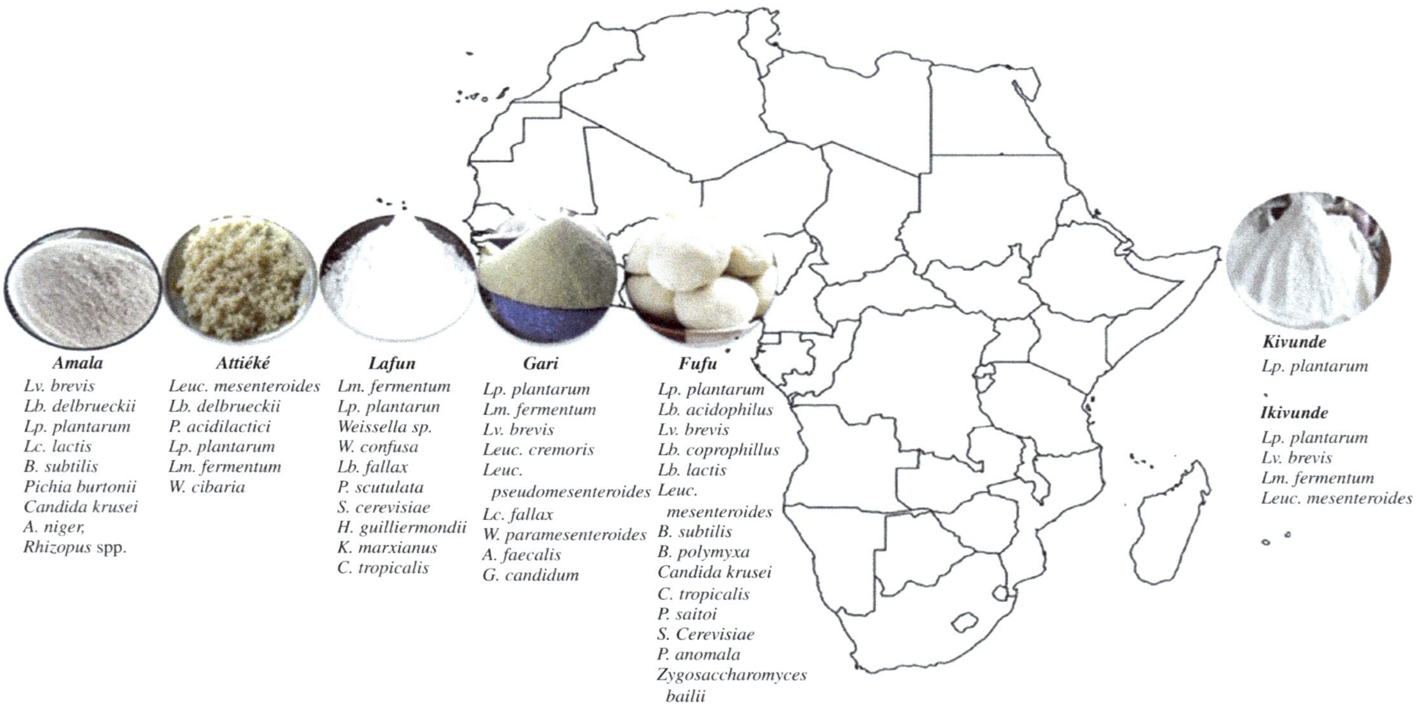

Figure 5.2 A pictorial view of various African fermented roots and tubers products from different regions of Africa, with their distinct fermenting microorganisms listed.

5.2.3 Attiéké

Attiéké is a starchy fermented cassava product. Originally from the Ivory Coast, it is now a popular staple food in other West African countries such as Benin, Burkina Faso and Togo. It is a steamed couscous-like product with a slightly sour taste and whitish colour, often served as a side dish with various sauces, stews or grilled meats (Coulin et al., 2006). In the Ivory Coast, *attiéké* is produced by the use of an inoculum rather than using a spontaneous fermentation method. There are three main ethnic groups that produce this traditional dish (Adjoukrou, Alladjan and Ebrie), and each uses a different variation in their method of production (Djeni et al., 2015). In summary though, the inoculum production follows this simple process. Prior to the production of *attiéké*, a traditional starter/inoculum called *lidjrou* is prepared by cooking peeled whole fresh cassava tubers (2–3 kg) in boiling water (100°C) for 5–10 minutes. This is then cooled at ambient temperature (28–35°C) and packed in a bag such as an old rice sack or fishing nets which have already been used for previous fermentations. The bag or net is packed into a basket or box and left to undergo spontaneous fermentation at 28–35°C for two to three days (Coulin et al., 2006). LAB, *Bacillus* spp., yeasts, faecal Enterococci and moulds may play a role in the inoculum production (Assanvo et al., 2006), although a recent study has confirmed five LAB species to be present, namely, *Leuconostoc mesenteroides*, *Pediococcus acidilactici*, *Lactiplantibacillus plantarum*, *Limosilactobacillus fermentum* and *Weissella cibaria* (Djeni et al., 2015).

The production of *attiéké* involves several steps. First, cassava roots are peeled, cut into pieces, washed and grated. During grating the cassava mass is mixed with 10% of a traditionally prepared starter/inoculum (*lidjrou*) and about 0.1% palm oil (Coulin et al., 2006). The inoculated pulp is fermented overnight in covered bins at 28–35°C. After fermentation, the pulp is filled into bags and pressed for several hours. The pressed pulp is taken from the bags and squeezed through a sieve to obtain granules that are sun-dried and then cleaned to remove fibres and waste. The dried granules are steamed to produce *attiéké*, which is sold in small plastic bags as a ready-to-eat food (Coulin et al., 2006). *Leuconostoc mesenteroides* subsp. *mesenteroides*, *Ligilactobacillus salivarius*, *Lb. delbrueckii* subsp.

delbrueckii, *Lp. plantarum*, *Lm. fermentum*, *Lb. confusus*, *Weissella cibaria*, *W. confusa*, *Bacillus* spp. (*B. circulans*, *B. lentus*), Enterococci, yeasts and moulds are common microbes found in *attiéké* (Coulin et al., 2006). The presence of coliforms such as *Enterobacter sakazakii*, *Ent. cloacae* and *Klebsiella pneumoniae* subsp. *pneumoniae* may indicate contamination from the environment and producers during production (Assanvo et al., 2006; Coulin et al., 2006; Djeni et al., 2015; Krabi et al., 2016). The bitterness of the cassava is eliminated during the long fermentation process (Assanvo et al., 2006; Fortin et al., 1998). Fermentation enhances the reduction of the cyanogenic glucosides present in cassava and improves the flavour and texture of the final product. The microbiology of *attiéké* is a fascinating interplay of various microorganisms, primarily LAB and yeasts, which work together to transform cassava into a safe, nutritious and flavourful product. The traditional fermentation process ensures the development of unique sensory characteristics and enhances the nutritional quality of *attiéké*, while reducing toxins.

5.2.4 Lafun

Lafun is a traditional West African fermented cassava product commonly consumed in Benin and Nigeria, where it is a staple food. The preparation of *lafun* begins with soaking peeled cassava chunks submerged in water at ambient temperature (28–32°C) for three to five days to undergo spontaneous fermentation. During this time, various microorganisms, particularly LAB and yeasts, play a crucial role in the fermentation process (Oyewole and Odunfa, 1988; Padonou, Nielsen, et al., 2009). The cassava chunks are later sun-dried and milled into a fine flour called *lafun* (Figure 5.1). The flour is used to cook a stiff porridge, often consumed with various soups and stews.

Different biochemical changes occur at the fermentation stage, such as degradation of cyanogenic compounds, formation of flavour compounds and retting/softening of the roots (Ampe & Brauman, 1995; Oyewole and Odunfa, 1988). The degree of root softening is the most essential criterion which determines the end of *lafun* fermentation (Oyewole and Odunfa, 1988; Padonou, Nielsen, et al., 2009). Consequently, the activities of pectinolytic enzymes such as pectin methylesterase and pectate lyase produced by the microorganisms are crucial during this stage of cassava fermentation (Ampe & Brauman, 1995; Okafor et al., 1984). The microbiota identified include several yeast species, with a predominance of *Saccharomyces cerevisiae*, *Pichia scutulata* and *Kluyveromyces marxianus*, LAB such as *Limosilactobacillus fermentum* and *Lactiplantibacillus plantarum* and spore formers, mainly *Bacillus cereus* (Adesemoye et al., 2025; Padonou, Nielsen, et al., 2009). The microbial succession during *lafun* fermentation in general indicates that the beginning of the process is a mixed microflora dominated by *B. cereus*, *Lp. plantarum* and *W. confuse*. At the same time, yeasts such as *K. pneumoniae* and *P. agglomerans* occur at low levels. After 24 hours of fermentation time, the microflora comprises *Lm. fermentum*, *B. cereus*, *W. confusa* and *Lp. plantarum* as the predominant bacteria and *K. marxianus* as the predominant yeast species. At 48 hours of fermentation time, the *Bacillus* species and Gram-negative bacteria tend to disappear due to the increase in acidification caused by the LAB, with the microbial population consisting of *Lm. fermentum*, *Lp. plantarum*, *Weissella* sp., *W. confusa*, *Lb. fallax*, *P. scutulata*, *Saccharomyces cerevisiae*, *H. guilliermondii*, *K. marxianus* and *Candida tropicalis* (Adesemoye et al., 2025; Padonou, Nielsen, et al., 2009). The yeasts, the LAB and *B. cereus* at the beginning of the spontaneous fermentation have been detected at levels of 3 log10, 5 log10 and 5–6 log10 CFU/g, respectively. The dominance of yeasts and LAB at the end of the fermentation process have been reported reaching levels of 5–6 log10 and 9 log10 CFU/g, respectively, while *B. cereus* is not detected at this point (Padonou, Nielsen, et al., 2009).

5.2.5 *Kivunde*

Kivunde is a traditional fermented cassava product consumed in Tanzania, usually produced using bitter cassava. Traditionally, whole cassava roots are washed with water to remove all dirt, then soaked in a stream of stagnant water underground for a week. By the end of the process, the fermented roots obtained have a smooth texture and a pleasant fruity aroma (Lorri & Svanberg, 1995). The controlled production of *kivunde* involves the peeling of fresh cassava, chopping, washing, submerging in water and soaking for three to five days for fermentation to take place, followed by washing, kneading and sun drying, and the final product (*kivunde*) is milled into fine powder

(Kimaryo et al., 2000). The microflora of *kivunde* is predominantly driven by *Lactiplantibacillus plantarum* (Kimaryo et al., 2000). In controlled fermentation using *Lactiplantibacillus plantarum* strains, the *kivunde* produced improvements in terms of product quality and lower levels of cyanide. *Lactiplantibacillus plantarum* strains improved fermentation based on their enzymatic activities and acid production (Ray & Sivakumar, 2009). Fermentation improves the nutritional profile of *kivunde* by increasing the bioavailability of nutrients, reducing antinutritional factors and enhancing the digestibility of the cassava.

5.2.6 Inyange and Ikivunde

The traditional African fermented cassava products of *inyange* and *ikivunde* are both popular staples in Burundi. In the production of inyange, fresh cassava roots are peeled, washed, cut into pieces and sun-dried for a day. They are then heaped together (heap fermentation) and covered with plantain leaves or straw, allowing them to ferment for five days. The fermentation period ends when moulds appear on the cassava's surface and the roots become soft, which contributes to the development of a desirable flavour. After fermentation, the moulds are scraped off and the cassava pieces are sun-dried, pounded in a wooden mortar and sieved to produce a slightly dark-coloured flour known as *inyange* (Hahn, 1989; Nzigamasabo & Hui Ming, 2006a, 2006b; Nzigamasabo & Nimpagaritse, 2009). The resulting product is cooked into a paste (*ubugali*) by stirring it with boiling water over low heat (Nzigamasabo & Nimpagaritse, 2009), typically served with legumes and meats. The production steps of *inyange* are summarised in Figure 5.1. Moulds such as *Aspergillus oryzae*, *Aspergillus fumigatus*, *Penicillium citrinum*, *Penicillium chrysogenum*, *Rhizopus stolonifer* and *Mucor* spp. are the predominant microorganisms in this SSF and are responsible for the colour and flavour of the flour. The yeast *Geotrichum candidum* is also present during the initial hours of fermentation (Nzigamasabo & Nimpagaritse, 2009).

In the preparation of *ikivunde*, cassava roots are peeled, washed, cut into pieces and soaked in an earthenware vessel for five days until they soften or froth appears on the surface of the water (Nzigamasabo & Nimpagaritse, 2009). The fermented roots are then sun-dried on mats, racks or rooftops for three days to a week. Once dried, the cassava pieces are pounded and sieved to produce a white flour known as *ikivunde* (Figure 5.1). This submerged fermentation technique is particularly used for bitter cassava, reducing its cyanide content and efficiently detoxifying it. The fermentation process, primarily driven by LAB such as *Lactiplantibacillus plantarum*, *Levilactobacillus brevis*, *Limosilactobacillus fermentum* and *Leuconostoc mesenteroides*, with *Lp. plantarum* being predominant, also involves the yeast *Geotrichum candidum* (Hahn, 1989; Nzigamasabo & Hui Ming, 2006a, 2006b; Nzigamasabo & Nimpagaritse, 2009). Fermentation reduces antinutritional factors such as cyanide and phytic acid, enhancing the product's safety and nutritional value.

5.2.7 Elubo isu or *Amala*

Elubo isu or *amala* is a traditional Nigerian food from fermented yam tubers. The preparation starts with washing, peeling and slicing the fresh yams, which are then left to dry on rocks or cemented floors for many days until a dried crust is formed, which is then milled into flour. *Amala* is the cooked form of this flour, prepared by mixing *elubo isu* with hot water to form a smooth, elastic dough.

The microbiology of *elubo isu* and *amala* is influenced by the microbial communities that develop during the drying and fermentation stages of preparation. *Elubo isu*, being a dried product, has relatively low moisture content, which limits the growth of many microorganisms. However, the initial stages of preparation, such as peeling and slicing, can introduce various microorganisms from the environment, the handling equipment and the raw materials themselves. The key microbial groups of bacteria involved in yam fermentation to produce *elubo isu* flour for making amala are *L. brevis*, *Lactobacillus delbrueckii*, *Lp. plantarum*, *Lactococcus lactis* and *B. subtilis*, while *Pichia burtonii* and *Candida krusei* are the predominant yeasts (Achi & Akubor, 2000; Osho et al., 2010). The presence of moulds such as *Aspergillus niger*, *Aspergillus flavus*, *Rhizopus* spp. and *Neurospora* spp. has been reported during the drying of yam chips (Adesemoye et al., 2025; Ayoade et al., 2018).

The microbiology of *elubo isu* involves various microorganisms that play roles during the drying and fermentation processes. Proper handling, drying and storage practices are essential to ensure the safety and quality of these traditional Nigerian foods. The fermentation process enhances the flavour and nutritional value of *elubo isu* and contributes to its extended shelf life.

5.3 Conclusion

The microbiology of African fermented root and tuber products showcases the rich diversity of microorganisms involved in traditional fermentation processes. Fermented cassava products such as *gari*, *fufu* and *kivunde* are staples in many African diets. These products are common in East and West Africa primarily because the crop thrives in these regions, making it a readily available and culturally significant food source. The fermentation process for cassava typically involves a combination of LAB and yeasts. The dominant LAB species include *Lactiplantibacillus plantarum*, *Limosilactobacillus fermentum* and *Lactococcus lactis*, which contribute to the sour taste, detoxification of raw cassava and improved product. Similarly, fermented yam products such as *amala* depend on microbial fermentation to enhance their digestibility and nutritional value. LAB species such as *Levilactobacillus brevis* and *Lactiplantibacillus plantarum* are commonly involved, contributing to the acidification process that helps preserve the yam and improve its flavour. The interactions between these microorganisms are critical in determining the final characteristics of fermented root and tuber products. LAB play a significant role in acid production, which lowers the pH and inhibits the growth of spoilage organisms and pathogens. Yeasts contribute to the development of flavour and aroma compounds, enhancing the sensory qualities of the products. The microbiology of African fermented root and tuber products is characterised by a diverse array of microorganisms that work synergistically to improve these foods' safety, nutritional value and sensory properties. The traditional fermentation processes harness natural microbial consortia, ensuring these essential food items' sustainability and cultural significance in African communities. Understanding and optimising these microbial interactions can further enhance the quality and health benefits of fermented root and tuber products, promoting food security and nutritional well-being across the continent.

References

Achi, O., & Akubor, P. (2000). Microbiological characterization of yam fermentation for 'Elubo' (yam flour) production. *World Journal of Microbiology and Biotechnology*, *16*(1), 3–7.

Adegoke, G. O., & Babalola, A. K. (1988). Characteristics of micro-organisms of importance in the fermentation of fufu and ogi—Two Nigerian foods. *Journal of Applied Microbiology*, *65*(6), 449–453.

Adesemoye, E. T., Sanni, A. I., Spano, G., Capozzi, V., & Fragasso, M. (2025). Lactic acid bacteria diversity in fermented foods as potential bio-resources contributing to alleviate malnutrition in developing countries: Nigeria as a case study. *Fermentation*, *11*(2), 103.

Adesulu, A. T., & Awojobi, K. O. (2014). Enhancing sustainable development through indigenous fermented food products in Nigeria. *African Journal of Microbiology Research*, *8*(12), 1338–1343.

Adeyemo, S., Agun, T., & Ogunlusi, E. (2018). Antimicrobial activity of lactic acid bacteria isolated from 'Pupuru': An African fermented staple against food borne-pathogens. *Journal of Molecular Biology and Biotechnology*, *3*(1), 5.

Adu-Gyamfi, C. A. (2022). An evaluation of glycaemic load in the assortments of fufu in Ghana. *Universal Journal of Food Science and Technology*, *1*(1), 12–32.

Aguoru, C. U., Onda, O., Omoni, V. T., & Ogbonna, I. O. (2014). Characterization of moulds associated with processed garri stored for 40 days at ambient temperature in Makurdi, Nigeria. *African Journal of Biotechnology*, *13*(5), 673–677.

Ahaotu, I., Ogueke, C. C., Owuamanam, C. I., Ahaotu, N. N., & Nwosu, J. N. (2013). Fermentation of undewatered cassava pulp by linamarase producing microorganisms: Effect on nutritional composition and residual cyanide. *American Journal of Food and Nutrition*, *3*(1), 1–8.

Ajayi, O. I., Ehiwuogu-Onyibe, J., Oluwole, O. B., Jegede, A. A., Salami, T. A., Asieba, G. O., Chiedu, I. E., Suberu, Y. L., Aba, E. M., Dike, E. N., Ajuebor, F. N., & Elemo, G. N. (2016). Production of fermented sweet potato flour using indigenous starter cultures. *African Journal of Microbiology Research*, *10*(41), 1746–1758.

Akindahunsi, A. A., & Oboh, G. (2003). Effect of fungi fermentation on organoleptic properties, energy content and in-vitro multienzyme digestibility of cassava products (flour & gari). *Nutrition and Health*, *17*(2), 131–138.

Akintayo, O. A., Dudu, O. E., Awoyale, W., Gerrano, A. S., Odunlade, T. V., Njobeh, P. B., & Oyeyinka, S. A. (2023). African Fermented root and tuber-based products. In *Indigenous fermented foods for the tropics* (pp. 265–283). Academic Press.

Amoa-Awua, W. K., Frisvad, J. C., Sefa-Dedeh, S., & Jakobsen, M. (1997). The contribution of moulds and yeasts to the fermentation of 'agbelima' cassava dough. *Journal of Applied Microbiology*, *83*(3), 288–296.

Amoa-Awua, W. K. A., Appoh, F. E., & Jakobsen, M. (1996). Lactic acid fermentation of cassava dough into agbelima. *International Journal of Food Microbiology*, *31*(13), 87–98.

Ampe, F., & Brauman, A. (1995). Origin of enzymes involved in detoxification and root softening during cassava retting. *World Journal of Microbiology and Biotechnology*, *11*, 178–182.

Assanvo, J. B., Agbo, G. N., Behi, Y. E. N., Coulin, P., & Farah, Z. (2006). Microflora of traditional starter made from cassava for "attiéké" production in Dabou (Côte d'Ivoire). *Food Control*, *17*(1), 37–41.

Ayoade, F., Adeniji, P., Amole, K., Amaremo, Y., Apata, T., Fayemi, S., Oyejide, N., Abazuh, U. D., Kayode, T., Daramola, G. G., & Folarin, O. (2018). The predominant lactic acid microorganisms of spontaneously fermented Amala, a Yam Food Product. *Asian Food Science Journal*, *4*(3), 1–10.

Babatuyi, C. Y., Boboye, B. E., Fagbemi, T. N., & Enujiugha, V. N. (2020). Cyanide, haematology and histopathology profiles of albino rats fed with 'Fufu'-based diets produced from mixed starter cultures. *Heliyon*, *6*(7).

Behera, S. S., & Ray, R. C. (2017). Microbial linamarase in cassava fermentation. In *Microbial enzyme technology in food applications* (pp. 333–346). CRC Press.

Braide, W., Azuwike, C. O., & Adeleye, S. A. (2018). The role of microorganisms in the production of some indigenous fermented foods in Nigeria. *International Journal of Advanced Research in Biological Sciences*, *5*(5), 86–94.

Coulin, P., Farah, Z., Assanvo, J., Spillmann, H., & Puhan, Z. (2006). Characterisation of the microflora of attiéké, a fermented cassava product, during traditional small-scale preparation. *International Journal of Food Microbiology*, *106*(2), 131–136.

Djeni, N. T., Bouatenin, K. P., Assohoun, N. M. C., Toka, D. M., Menan, E. H., Dousset, X., & Dje, K. M. (2015). Biochemical and microbial characterization of cassava inocula from the three main attieke production zones in Côte d'Ivoire. *Food Control*, *50*, 133–140.

Ferraro, V., Piccirillo, C., Tomlins, K., & Pintado, M. E. (2016). Cassava (*Manihot esculenta* Crantz) and yam (*Dioscorea* spp.) crops and their derived foodstuffs: Safety, security and nutritional value. *Critical Reviews in Food Science and Nutrition*, *56*(16), 2714–2727.

Fortin, J., Desmarais, G., Asovié, O., & Diallo, M., (1998). Attiéké, couscous of Ivory Coast. In Latour, J. (ed.), *Horizon francophonie* (pp. 22–24). The Food World.

Hahn, S. K. (1989). An overview of African traditional cassava processing and utilization. *Outlook Agriculture*, *18*, 110–118.

Ihenetu, F. C., Ihenetu, S. C., Ohalete, C. N., Njoku-Obi, T. N., & Okereke, B. C. (2017). Microorganism associated with cassava fermentation for Abacha product sold in Owerri, Imo State. *World News of Natural Sciences*, *12*, 4350.

Kimaryo, V. M., Massawe, G. A., Olasupo, N. A., & Holzapfel, W. H. (2000). The use of starter culture in the fermentation of cassava for the production of "kivunde," a traditional Tanzanian food product. *International Journal of Food Microbiology*, *56*, 179–190.

Kostinek, M., Specht, I., Edward, V. A., Schillinger, U., Hertel, C., Holzapfel, W. H., & Franz, C. M. (2005). Diversity and technological properties of predominant lactic acid bacteria from fermented cassava used for the preparation of gari, a traditional African food. *Systematic and Applied Microbiology*, *28*(6), 527–540.

Krabi, E. R., Assamoi, A. A., Ehon, A. F., Amani, N. G. G., Niamké, L. S., Cnockaert, M., Aerts, M., & Vandamme, P. (2016). Biochemical properties of three lactic acid bacteria strains isolated from traditional cassava starters used for attieke preparation. *African Journal of Food Science*, *10*(11), 271–277.

Lorri, W., & Svanberg, U. (1995). An overview of the use of fermented foods for child feeding in Tanzania. *Ecology of Food and Nutrition*, *34*, 65–681.

Nzigamasabo, A., & Hui Ming, Z. (2006a). Comparative study on nutrients and antinutrients changes in Ikivunde and Inyange, two Burundian traditionally processed cassava products. *Journal of the Science of Food and Agriculture*, *86*(12), 1878–1886.

Nzigamasabo, A., & Hui Ming, Z. (2006b). Traditional cassava foods in Burundi – A review. *Food Reviews International*, *22*(1), 1–27.

Nzigamasabo, A., & Nimpagaritse, A. (2009). Traditional fermented foods and beverages in Burundi. *Food Research International*, *42*(5–6), 588–594.

Oboh, G., & Akindahunsi, A. A. (2003). Biochemical changes in cassava products (flour & gari) subjected to *Saccharomyces cerevisae* solid media fermentation. *Food Chemistry*, *82*(4), 599–602.

Ogbonnaya, N. (2016). Linamarase production by some microbial isolates and a comparison of the rate of degradation of cassava cyanide by microbial and cassava linamarases. *Hemijska industrija*, *70*(2), 129–136.

Oguntoyinbo, F. A. (2007). Identification and functional properties of dominant lactic acid bacteria isolated at different stages of solid-state fermentation of cassava during traditional gari production. *World Journal of Microbiology and Biotechnology*, *23*, 1425–1432.

Oguntoyinbo, F. A., & Dodd, C. E. (2010). Bacterial dynamics during the spontaneous fermentation of cassava dough in gari production. *Food Control*, *21*(3), 306–312.

Okafor, N., Ijioma, B., & Oyolu, C. (1984). Studies on the microbiology of cassava retting for foo-foo production. *Journal of Applied Bacteriology*, *56*(1), 1–13.

Osho, A., Mabekoje, O. O., & Bello, O. O. (2010). Comparative study on the microbial load of Gari, Elubo-isu and Iru in Nigeria. *African Journal of Food Science*, *4*(10), 646–649.

Oyedeji, O., Ogunbanwo, S. T., & Onilude, A. A. (2013). Predominant lactic acid bacteria involved in the traditional fermentation of fufu and ogi, two Nigerian fermented food products. *Food and Nutrition Sciences*, *4*(11), 40.

Oyewole, O. B. (1995). Application of biotechnology to cassava processing in Africa. In Egbe, T.A., Brauman, A., Griffon, D., & Treche, S. (Eds.), *Transformation Alimentaire Du Manioc (Cassava Food Processing)* (pp. 277–286). ORSTROM.

Oyewole, O. B. (2001). Characteristics and significance of yeasts' involvement in cassava fermentation for 'fufu'production. *International Journal of Food Microbiology*, *65*(3), 213–218.

Oyewole, O. B., & Odunfa, S. A. (1988). Microbiological studies on cassava fermentation for 'lafun' production. *Food Microbiology*, *5*(3), 125–133.

Oyewole, O. B., & Odunfa, S. A. (1990). Characterization and distribution of lactic acid bacteria in cassava fermentation during fufu production. *Journal of Applied Bacteriology*, *68*(2), 145–152.

Oyewole, O. B., & Odunfa, S. A. (1992). Extracellular enzyme activities during cassava fermentation for 'fufu' production. *World Journal of Microbiology and Biotechnology*, *8*, 71–72.

Padonou, S. W., Hounhouigan, D. J., & Nago, C. M. (2009). Physical, chemical and microbiological characteristics of lafun produced in Benin. *African Journal of Biotechnology*, *8*, 3320–3325.

Padonou, S. W., Nielsen, D. S., Hounhouigan, D. J., Thorsen, L., Nago, C. M., & Jakobsen, M. (2009). The microbiota of Lafun, an African traditional cassava food product. *International Journal of Food Microbiology*, *133*, 22–30.

Panda, S. K., & Ray, R. C. (2016). Fermented foods and beverages from tropical roots and tubers. In *Tropical roots and tubers: Production, processing and technology* (pp. 225–252). Wiley.

Panghal, A., Munezero, C., Sharma, P., & Chhikara, N. (2019). Cassava toxicity, detoxification and its food applications: A review. *Toxin Reviews*, *40*(1), 1–16.

Ray, R. C., & Sivakumar, P. S. (2009). Traditional and novel fermented foods and beverages from tropical root and tuber crops. *International Journal of Food Science & Technology*, *44*(6), 1073–1087.

Ray, R. C., & Ward, O. P. (2006). Postharvest microbial biotechnology of tropical root and tuber crops. In *Microbial biotechnology in horticulture* (Vol. 1, pp. 345–396). CRC Press.

Sanni, L. O., Ikuomola, D. P., & Sanni, S. A. (2001). Effect of length of fermentation and varieties on the qualities of sweet potato gari. In Akoroda, M. O. (Ed.), *Proc. 8th triennial Symposium of the International Society for Tropical Root Crops* (pp. 208–211). Africa Branch (ISTRC-AB), IITA, Ibadan, Nigeria.

Tsav-Wua, J., Inyang, C., & Akpapunam, M. (2004). Microbiological quality of fermented cassava flour 'kpor umilin. *International Journal of Food Sciences and Nutrition*, *55*(4), 317–324.

Wakil, S., & Benjamin, I. (2015). Starter developed pupuru, a traditional Africa fermented food from cassava (*Manihot esculenta*). *International Food Research Journal*, *22*(6), 2565–2570.

6

Microbiology of African Fermented Fish and Meat-based Products

6.1 Introduction

Many nations worldwide incorporate fermented meat and fish products as a traditional part of their diets. These can be consumed as a side dish, sauce or main meal. The products are not only popular due to their distinct flavour, texture and high nutritional value but also due to the simplicity of their production methods, which can vary by nation. Historically, fermented meat and fish products have been produced using ancient methods passed through generations. Variations in the documented procedures and components indicate that these methods were typically applied without scientific knowledge or technological resources. Consequently, the sensory benefits were primarily based on changes in colour, odour and taste brought about by physical, chemical and microbiological changes not understood at the time these techniques were developed (Anihouvi et al., 2012; Bamidele et al., 2023; Belleggia & Osimani, 2023).

Fermented meat and fish products are essential sources of minerals and vitamins for numerous people, in addition to providing a substantial amount of protein (Tamang et al., 2020). For generations, Africans have embraced meat and fish fermentation, a practice rooted in tradition. This preservation method arose from the need to prevent spoilage in hot climates, where both meat and fish deteriorate quickly (Joardder & Masud, 2019). During fermentation, metabolic activity can modify the nutritional and bioactive properties of meat and fish, potentially leading to positive effects on human health. Consuming fermented meat and fish might yield antioxidant, antihypertensive, anticoagulant and anticancer benefits (Ashraf et al., 2020).

DOI: 10.1201/9781003658320-6

The demand for these products is growing worldwide due to the unique flavour and preservation qualities of traditionally or locally fermented products (Ashraf et al., 2020).

Meat and fish fermentation procedures largely rely on a combination of certain key steps, including salting, removing fats, adding spices or additives and maintaining aerobic or anaerobic conditions (to select beneficial microorganisms). They can also undergo mechanical treatment by being sliced, pressed, gutted, dried and/or smoked prior to fermentation (Bamidele et al., 2023; Oetterer et al., 2003). Salting is one of the oldest and most efficient ways to preserve meat and fish. In particular, adding salt can cause numerous physicochemical changes in fish and meat tissues, such as decreased water activity and moisture content and increased solubility of denatured proteins (Belleggia & Osimani, 2023; Nahar et al., 2017).

In the fermentation practices of these products, microorganisms including bacteria, yeasts and moulds initiate biochemical transformations crucial for nutrient preservation and product development. Lactic acid bacteria (LAB), notably species from the *Lactobacillus* and *Pediococcus* genera, are primary fermenters responsible for acidification, which enhances the sensory attributes of a product and inhibits the growth of spoilage and pathogenic microorganisms (Ghosh et al., 2023). Microbial populations that contribute to the fermentation process may vary significantly depending on factors such as the type of meat or fish, the fermentation method employed and the stage of fermentation. Some microorganisms may dominate under specific conditions, but the lack of comprehensive, region-specific studies means that these microbial dynamics in African meat and fish products are not well characterised. Therefore, a thorough comprehension of the microorganisms that thrive in African fermented meat and fish products is essential.

6.2 Fermented Meat Products in Africa

Fermented meat products are central to many African cultures, offering distinct tastes, textures and preservation techniques. Before being subjected to controlled fermentation, these traditional foods are prepared by treating the raw meat with a blend of salt, spices and, occasionally, acidic elements such as vinegar or fermented plant extracts.

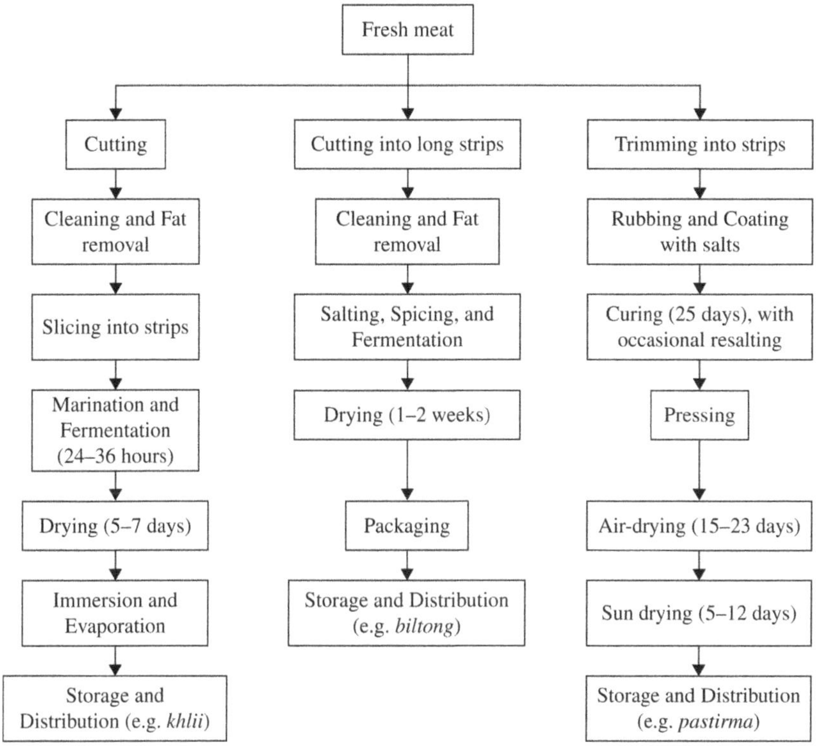

Figure 6.1 General traditional method of African meat fermentation.

This fermentative process encourages the growth of beneficial bacteria, yeasts or moulds, which enzymatically alter the proteins and sugars in meat. These products have been integral to cultural and nutritional practices across the continent for centuries, with diverse nations nurturing their own unique product names and preparation methods. This chapter details notable examples from various African regions and Figure 6.1 provides an overview of the general traditional methods of fermenting African meat products.

6.2.1 Khlii/khlia

Khlii, a delicacy originating from Morocco, is prepared from salted-dried beef, lamb, goat or camel meat (Figure 6.2) (Gagaoua & Boudechicha, 2018). The meat is cleaned and the fat is removed. It is sliced precisely into strips measuring 510 cm wide, 40 cm long and 2 cm thick and then marinated in a pre-prepared mixture of salt, crushed garlic,

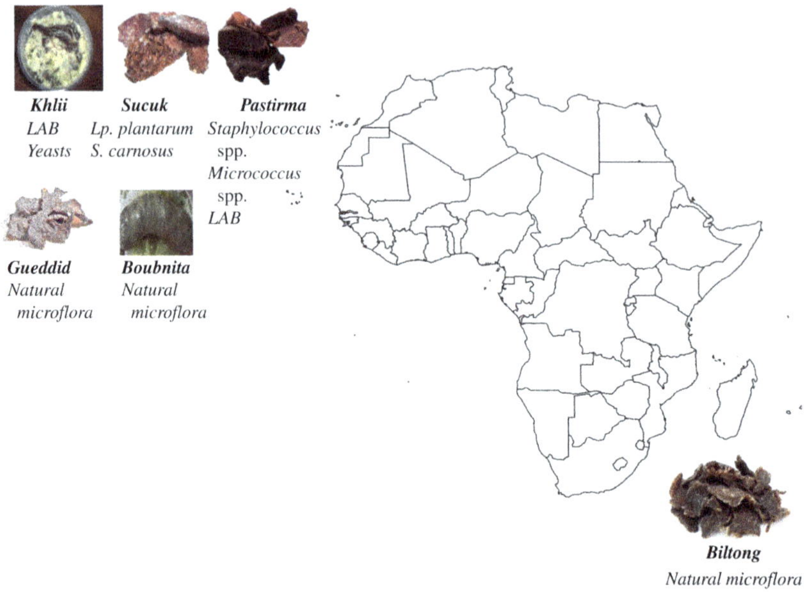

Figure 6.2 A pictorial view of various African fermented meat products from different regions of Africa and the distinct fermenting microorganisms. (*Sucuk*, *pastirma* (From Bamidele et al., 2023), *khlii* (From Benkerroum, 2013), *gueddid*, *Boubnita* (From Gagaoua & Boudechicha, 2018), *biltong* (https://picryl.com/media/africa-african-beef-food-drink-f34575).)

cumin, ground coriander grains, vinegar and seed oil, called *sharmula*. The marinated meat strips are then stored in a cool environment for 24–36 hours (fermentation) with intermittent stirring. Subsequently, the cured beef strips are left to sun-dry for five to seven days, hung on a laundry line or wire (Gagaoua & Boudechicha, 2018).

Once dried, these meat strips are immersed in liquid fat, cooked and conditioned in either plastic or glass containers until complete moisture evaporation occurs (Gagaoua & Boudechicha, 2018). Properly prepared and conditioned, *khlii* can be stored at room temperature for over two years, serving as a convenient ready-to-eat beef product or a delectable addition to breakfast when fried with eggs. In addition, it is a versatile pizza topping and an ingredient in various traditional dishes (Gagaoua & Boudechicha, 2018).

6.2.2 Soudjouk/Sucuk

Sucuk, depicted in Figure 6.2, is a culinary creation utilising lamb meat, beef, camel or water buffalo. Its production and consumption have been ingrained in North African cultures for numerous generations

(Gagaoua & Boudechicha, 2018). Traditionally, *sucuk* is made from a blend of ground beef, sheep tail fat (comprising 90% lean meat and 10% fat), salt, sugar, garlic and an array of spices such as paprika, black or red pepper and cumin. This mixture is then wrapped in natural casings, often fashioned from sheep's small intestines. Subsequently, the sausage is suspended for several weeks to undergo maturation and drying at room temperature (20–25°C) and controlled humidity until reaching a water activity level of less than 0.90, which contributes to its characteristic firm texture.

Sucuk fermentation can occur naturally through indigenous microorganisms or by incorporating starters, predominantly to expedite the process and ensure consistency in the final product's quality (Ercoskun et al., 2010). Optimal fermentation and flavour development necessitate regular moisture release from the product (Ercoskun et al., 2010). With a final pH ranging from 4.8 to 5.5 and moisture content between 4.2% and 6.3%, *sucuk* is typically sliced and enjoyed as a breakfast accompaniment (paired with eggs or cheese) or utilised as a filling for sandwiches (Ercoskun et al., 2010). It is also a much enjoyed spread for savoury pastries and sandwiches, as well as an appetiser (Ercoskun et al., 2010).

6.2.3 Pastirma

Pastirma, sometimes referred to as *pastrami*, *basterma* or *basturma*, is a common ready-to-eat meat delicacy in Egypt. Its name derives from the Turkish word bastrma, which means 'pressing', highlighting how integral pressing is to its preparation (Figure 6.2) (Lücke, 2017). Three key steps drive the weeks-long traditional process of producing *pastirma*; pressing, salting and then drying and ripening (Lücke, 2017). It is made with a range of lamb, camel and whole cow or water buffalo muscle cuts, and the choice of muscle slices significantly influences the quality of the final product.

The first step is to cut the meat into strips 60 cm long and 5 cm wide. Next, these strips are stacked, rubbed thoroughly and coated with salt before being allowed to cure for 25 days. Occasional turning or resalting of the meat is possible during this curing stage. Once the excess salt has been removed, the salted strips are air-dried for a maximum of 15 days in winter and around 23 days in summer (Benkerroum,

2013). Before being subjected to additional drying, the dried blocks are usually crushed severely. After the pressing and drying steps, the entire surface is then covered with a paste called *c'emen*, which is made of fenugreek, garlic and spicy pepper. Ultimately, it takes 5–12 days of sun drying to reach a level of readiness. A final pH of 5.7–6.1 is typically obtained with *pastirma* (Lücke, 2017). The most common microbes associated with *pastirma* are *Staphylococcus, Micrococcus* and LAB (Gagaoua & Boudechicha, 2018). *Pastirma* slices are typically prepared and gently roasted over a charcoal fire and eaten with scrambled eggs (Lücke, 2017). However, more recently, they have been used in cuisine as pizza toppings.

6.2.4 Biltong/Mukoki/Muhwaba

Various processes, such as salting, seasoning, fermentation and drying, are involved in making *biltong* (Figure 6.2), a well-known, ready-to-eat, salted and dried meat product from South Africa (Jones et al., 2017). This dried meat is also known as *mukoki* or *muhwaba* in Venda, South Africa. The meat is cut into long strips with a width/thickness of 2.5–5 cm, depending on muscle type, personal preference and drying period (Jones et al., 2017). *Biltong* is salted and coated with spices and sometimes vinegar, using several different techniques. The seasoned and salted meat strips are then left to ferment and dry. Traditionally, storage was at ambient temperature but nowadays can vary between 4 and 8°C for 18 and 20 hours (Burfoot et al., 2010; Naidoo & Lindsay, 2010). *Biltong* is often dried by hanging the beef strips outside for one to two weeks, depending on the atmospheric conditions. The meat strips are suspended on wire or plastic hooks in shaded areas, predominantly during the winter months when the wind, humidity and temperature are considered optimal. Due to the increasing demand for *biltong*, these traditional methods have been modified to achieve commercial profitability (Van der Riet, 1982).

Depending on the type being sold (sliced, sticks, slabs, etc.), biltong is either packed or stored in over-wrapped trays or loose paper bags. When preparing biltong at home, consumers also keep it in closed plastic containers or brown bags. However, to extend their shelf life, industrially produced biltong is supplied in packaging with a contained environment (Strydom & Zondagh, 2014). The storage conditions of biltong are crucial when hung outside, as these conditions can impact the colour, texture, moisture content and microbiological stability of the product. Because traditional biltong has low water activity (a_w), pH levels between 4.8 and 5.8 and curing salts, it is generally considered microbially stable (Petit et al., 2014). However, the issue of whether or not microbes impact the ultimate quality of the *biltong* product has not been thoroughly studied. Todorov et al. (2007) only assessed biltong as an unfermented meat product made by salt-drying. As a result, the literature has little information regarding the existence and role of microorganisms in the transformation and quality of the finished biltong products.

6.2.5 *Boubnita*

Boubnita is a traditional Moroccan dry-fermented sausage typically prepared and consumed following the Islamic festival of Eid al-Adha. The term '*Boubnita*' originates from the Maghreb dialect, in which '*Boubnita*' denotes the large intestine of cattle or sheep used as sausage casing (Gagaoua & Boudechicha, 2018). The preparation involves dicing lamb meat into small cubes and mixing it with a blend of spices, including salt, cumin, ginger, paprika, hot red pepper and olive oil. This seasoned meat is then carefully stuffed into thoroughly cleaned lamb intestines. The filled casings are tied with string and suspended in shaded outdoor areas, where they undergo slow drying and natural fermentation (Benkerroum, 2013). Once fully matured, *Boubnita* is commonly served in a sauce accompanied by vegetables or pasta.

6.2.6 *Afo-nnama*

Afo-nnama is a type of fermented product consumed in the northern parts of Nigeria. The meat from a variety of animals, including pigs, cows, camels and goats, is cleaned and then marinated for a few minutes in a brine solution. It is then hung on a rope in an open area for 15 days to allow fermentation to take place, which produces a unique aroma. The fermented, dried meat, known as *afo-nnama*, is soaked in either cold or hot water overnight before cooking (Bamidele et al., 2023).

6.2.7 *Gueddid*

Gueddid is a traditional dried fermented meat product commonly prepared in Morocco, Algeria and Tunisia (Figure 6.2). The product is made without added starter cultures, relying instead on the meat's natural microflora. Typically prepared from lamb or beef, the meat is first chopped, washed and then left at room temperature for approximately two days to undergo mild fermentation (Gagaoua & Boudechicha, 2018). The lean meat is then cut into strips roughly 4 cm in width, 40–60 cm in length and 1–2 cm in thickness. These strips are heavily salted and exposed to direct sunlight for 5–7 days to achieve the desired level of dehydration. Prior to consumption, *Gueddid* is rehydrated and desalinated by soaking in water, after which it is incorporated into a variety of traditional dishes (Gagaoua & Boudechicha, 2018).

6.2.8 *Beirta, Miriss* and *Dodery*

In Sudan, traditional fermented meat products such as *Beirta*, *Miriss* and *Dodery* are commonly prepared and consumed. *Beirta* is made using a mixture of organ meats from various animals like cows, camels, pigs and goats (Dirar, 1994). These organ meats, along with caul fat, are chopped and blended with approximately 2 kg of minced hindquarter muscle meat in a cooking pot. After the addition of half a litre of milk, the mixture is left to ferment for four days. On the fourth day, the container is opened, a small amount of salt is mixed in, and the pot is sealed again to allow fermentation to proceed for an additional three days. In Sudanese cuisine, this fermented meat mixture is often used to flavour sauces served with porridge (Dirar, 1994). *Miriss* is produced by fermenting the caul fat that encases the animal's stomach. This fat is kneaded with *combu* and left to ferment for six days. The resulting product is intensely white and typically has a strong, unpleasant odour (Dirar, 1994). On the other hand, *Dodery* is made from bone ends rich in marrow. These bones are cut into pieces, sun-dried, ground into a paste and then mixed with *combu*. The mixture is fermented for five days to produce the final product (Dirar, 1994).

6.2.9 Suya

Suya is a well-known street food in Nigeria made with thinly sliced meat, usually chicken or beef, that is skewered and cooked over an open flame. *Suya* itself is not fermented but is sometimes consumed with a hot and spicy peanut sauce known as *yaji* or *suya* spice, which may contain fermented ingredients such as locust bean (*dawadawa*). Cutting the meat into thin slices or strips is the first stage in making *suya* products. The meat is then combined with a marinade, which usually contains ginger, garlic, onion powder, cayenne pepper, paprika, crushed peanuts (or peanut butter), salt and occasionally bouillon powder or other spices.

Next, for consistent penetration, the thinly sliced meat is covered with the marinade mixture for a few hours, preferably overnight. Once the meat has marinated, it is threaded onto skewers, either bamboo or metal. After that, the skewered meat is cooked over a charcoal or open flame. It is customary to stir it regularly and baste it with oil or a mixture of oil and spices to prevent the meat from drying out and to release the taste. When the meat is well done and acceptably charred on the outside, it is typically served hot with sliced tomatoes, onions and occasionally fresh pepper slices.

6.3 Microbiology of African Fermented Meat Products

Meat degradation is predominantly caused by the actions of most microorganisms, although fermentative microorganisms, particularly LAB, can be used to produce fermented meat products (Mokoena et al., 2016). They either can be added by manufacturers or occur naturally in raw meat (natural microflora). The desired microorganisms, known as starter cultures, may include a single species or a combination of microorganisms that are often added to meat products (Kandasamy et al., 2018). According to Koza et al. (2008), Gram-positive bacteria belonging to the genera *Lactobacillus*, *Micrococcus* and *Staphylococcus* play a crucial role in the fermentation and maturation of sausages. These microbial species have also been isolated in various types of fermented African meats (Zakpaa et al., 2009).

Other LABs less frequently detected including *Lactococci*, *Leuconostocs*, *Pediococci* and *Enterococci* have little to no effect on the majority of traditional naturally fermented sausages. Various types of catalase-positive cocci may be found during meat fermentation, depending on the raw ingredients used and processing circumstances (Leroy & De Vuyst, 2016). To maintain the colour stability and avoid rancidity of the cured meat, *Micrococcaceae* species are utilised to increase the quantity of fermentative bacteria throughout the ageing process of fermented meat products. The mechanism of action of this microbial group shortens processing times and promotes flavour and taste while preventing the growth of spoilage microbes (Fraqueza & Patarata, 2019). *Micrococci* are also thought to play a role in the beneficial processes of colour stabilisation, peroxide breakdown, lipolysis and proteolysis that take place during the ripening of dry-fermented sausages (Petrovic et al., 2025). Some research has shown that the presence of *Staphylococcus* species can change the smell of fermented meat products (Laranjo et al., 2019). According to a study on the respective contributions of LAB and *Micrococcaceae* to fragrance synthesis, *Staphylococcus carnosus* and *Staphylococcus xylosus* may both produce esters and other vital aromatic components. Furthermore, due to their high catalase and nitrate reductase activity, these strains inhibit the formation of rotten flavours (Carballo, 2012).

The microbe used as a starting culture can impact some qualitative characteristics of fermented meat products (Cruxen et al., 2019). The decline in pH levels brought on by LAB activity affects the colour of the product, as do the effects of *Micrococcaceae* on nitrate reduction, oxygen consumption and hydrogen peroxide breakdown (Kumar et al., 2017). In addition, these microbial activities are related to O_2 usage and H_2O_2 decomposition, and this influences the colour of products that use moulds and yeasts as starter cultures (Laranjo et al., 2019). The aroma of fermented meat products is another crucial feature affected by microbe activity. Moulds, yeasts and *Micrococcaceae* are very adept starter cultures due to their ability to break down proteins and lipids and delay rancidity (Leroy et al., 2006). The texture of meat products is exclusive to the LAB genera. On the other hand, the physical appearance and resistance to light, oxygen and drying of some products, such as *soudjouk* in Egypt, are directly attributed to either mould and/or yeast (Gagaoua & Boudechicha, 2018). Textural or structural stability is one of the most significant quality indicators for beef products that have undergone fermentation. LAB and *Micrococcaceae* primarily affect product stability by lowering pH and nitrate levels and inhibit the growth of spoilage microorganisms (Xiao et al., 2018).

The activity of meat enzymes, especially proteases and lipases, causes the product to age and ripen during drying, just as with salted meat products. Unlike dry-salted hams, the initial process for producing dry-fermented sausages includes the spontaneous acidity of the meat matrix, which acts as a preservation hindrance in addition to the initial salting and drying. Salt-solubilised protein fractions are denatured to a distinctive sliceable and gel-like texture, and the growth of undesirable microbes present in raw meat is inhibited by fermentation acidification of pH levels between 4.8 and 5.2. The combination of curing salt and anaerobic conditions in the sausage casings at the appropriate temperature (18–28°C) significantly selects specific and beneficial microorganisms (Leroy & De Vuyst, 2016).

6.4 Fermented Fish Products in Africa

Many African consumers still prefer fresh fish even if artisanal methods are used to preserve a large amount of the landed produce. The most popular methods for preparing fish are grilling, frying, fermenting, smoking, salting and sun drying (El Sheikha et al., 2014). The type of fishery product available in any particular country is directly related to the dietary habits and purchasing power of the populace (Costello et al., 2020). Africa has several seas, rivers, lakes and oceans; these bodies of water are substantial sources of various fresh fish species for many people. Over 80% of fish harvests consumed in Africa occur in West Africa, specifically from the Atlantic Ocean. The remainder derives from freshwater resources, including the River Shari, Lake Volta, Lake Chad and the Niger River. These lakes produce abundant fish that are salted, pickled and dried for export and home consumption (Adeyeye & Oyewole, 2016).

Figure 6.3 depicts traditional fish fermenting techniques used in Africa. Although rooted in tradition, most processing techniques have advanced significantly to mitigate contamination and potential health hazards that may emerge throughout the processing steps.

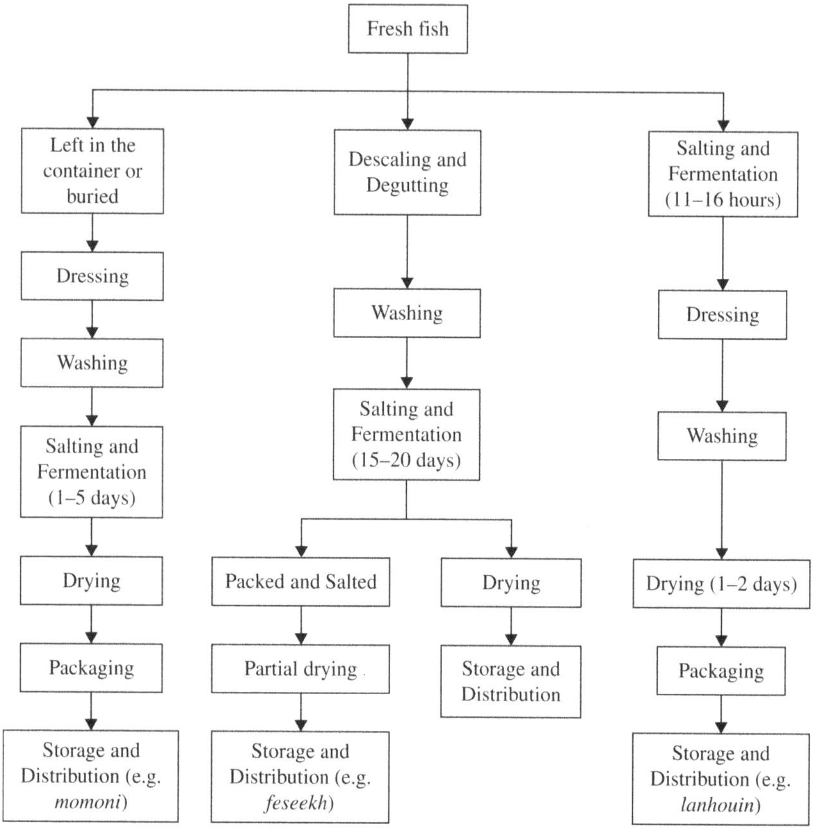

Figure 6.3 General traditional method of African fish fermentation. (Modified from Bamidele et al., 2023 and Essuman, 1992.)

The most notable African fermented fish products are covered in this section. Fermented fish products are less prevalent in Southern Africa than in other regions.

6.4.1 Lanhouin

Fish products can ferment naturally, which is a predominantly uncontrolled process. Before being cleaned and allowed to mature for 11–16 hours, fresh fish are gutted and scaled, and the gills removed (Anihouvi et al., 2006). The delicate dressed fish is then salted and fermented in a basket for 39 hours. The fermented fish is water rinsed and further sun-dried for 12 days. The lack of control over the fermentation processes and the final product (typically of varying quality)

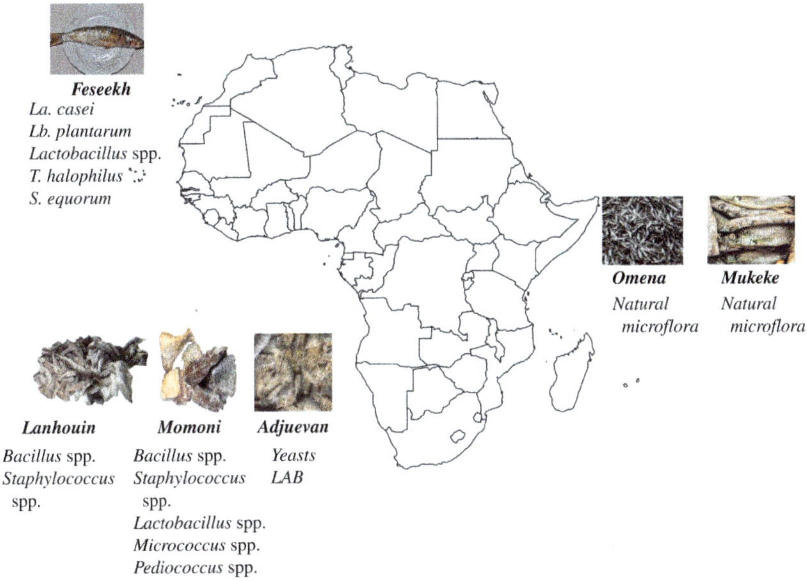

Figure 6.4 A pictorial view of various African fermented fish products from different regions of Africa and the distinct fermenting microorganisms. (*Lanhouin, momoni* (From Bamidele et al., 2023), *adjuevan* (From Kouassi et al., 2024), *omena* (From Marijani et al., 2020), *mukeke* (https://en.wikipedia.org/wiki/Sleek_lates), *feseekh* (https://en.wikipedia.org/wiki/Fesikh).)

increases the possibility of quality defects, which is a disadvantage of this fermentation method. Generally made from cassava or king-fish, *lanhouin* (Figure 6.4) is frequently used as a condiment in several African nations (Ghana, Togo and Benin).

6.4.2 Momoni

In Ghana, *momoni* (Figure 6.4) is a popular fermented fish product frequently used as a condiment when making stew for consumption with various foods, such as boiling unripe plantains, yam and cocoyam (El Sheikha et al., 2014). Various freshwater fish can be used to prepare *momoni* (Figure 6.4). The fish is cleaned and salted after being scaled and gutted. The parts of *momoni* attached to the gills and stomach are heavily salted. According to El Sheikha et al. (2014), the seasoned fish are put in pots or baskets which are covered with jute bags or aluminium trays to ferment for 15 days. The fermented fish is then rinsed with salted water, and the chopped, fermented fish pieces are sun-dried in an open area on a wooden tray for a few hours. *Momoni* is a solid consistency when added to a boiling stew containing tomato, onion, ground red pepper and a small quantity of palm oil. The product frequently has a high salt content, is of low quality and deteriorates rapidly in storage and during retail (El Sheikha et al., 2014).

6.4.3 *Adjuevan/Adjonfa*

The Ivorian traditional fermented fish product *adjuevan* (Figure 6.4) is made from Atlantic bumper, *Chloroscombrus chrysurus* using natural fermentation. On the west coast of Côte d'Ivoire, *adjuevan*, a salted and fermented fish product, is typically produced at room temperature (28–30°C) using two traditional methods (El Sheikha et al., 2014). The first method includes fermenting the product for five days in jars covered with plastic and stones, and the second technique entails fermenting the product for at least ten days before drying it on racks or nets (Kouakou et al., 2012). The strong smell of *adjonfa* makes it a popular condiment with different flavourings and cuisines for seasoning sauces for yam, plantain, *attieke* and other similar meals (Koffi-Nevry et al., 2011).

6.4.4 *Feseekh*

Feseekh is the ancient Egyptian word for the *Mugil cephalus* fish when salted and fermented (Figure 6.4) (El-Sebaiy & Metwalli, 1989). The Egyptian market offers two varieties: one with a low salt level that may be consumed after 15–20 days of maturation, and the other with a high salt content that can be consumed after two to three months of preservation. *Feseekh* is highly nutritious and can be a rich source of essential amino acids, vitamins, minerals and protein (Zang et al., 2020).

The process of making *feseekh* is complex, and in some families the method and the processing are handed down from one generation to the next. *Feseekh* is usually consumed at *Sham El Nesem* (Smelling the Breeze), an ancient Egyptian spring event.

6.4.5 *Mukeke*

Mukeke is a traditional fermented and dried fish product widely consumed in the coastal and freshwater regions of Tanzania and Uganda

(Figure 6.4). It is commonly prepared using small pelagic fish species such as Nile perch and silver cyprinid, which are heavily salted and allowed to ferment and dry naturally under ambient conditions over a period of three to seven days (Baniga et al., 2017). The salt concentration typically ranges between 10 and 20%, which prevents the proliferation of spoilage organisms while favouring the growth of halotolerant and fermentative microbes. The microbial ecology of *Mukeke* is not extensively studied, but its fermentation is generally spontaneous, relying on the natural microbiota associated with the fish and the surrounding environment.

6.4.6 *Bissara* with *Bakalao*

Bissara is a popular dish in Morocco, typically made from dried, puréed fava beans. Sometimes, especially in coastal regions, fermented fish such as *bakalao* (salted cod) is added to *bissara* to create a unique flavour profile. The fermented fish adds a salty and savoury element to the dish.

6.4.7 *Omena*

Also known as *dagaa* in Tanzania, *omena* is a small, silver-coloured fish that is sun-dried and then fermented (Figure 6.4). It is a common ingredient in East African cuisine, especially in Kenya and Tanzania, where it is often cooked with vegetables or served with *ugali* (maize meal).

6.4.8 *Dakkous*

Dakkous is a traditional fermented fish sauce made by fermenting fish with salt, sometimes mixed with spices like *fenugreek* for some time. It is commonly used as a condiment in Ethiopian cuisine.

6.4.9 *Oshifima* with *Omaheke Mopane Worms*

Oshifima is a traditional staple food in Namibia made from *mahangu* (pearl millet) flour. Sometimes, fermented fish is added to *oshifima* to enhance its flavour. In some regions of Namibia, dried *mopane* worms, known as *omaheke*, are fermented and then added to dishes such as *oshifima* for added protein and flavour.

6.5 Microbiology of African Fermented Fish Products

Fish have microflora in the slime on their bodies, guts and gills in their natural habitats. When a fish dies, putrefactive changes are caused by microorganisms and the enzymes in its tissues. In addition, the microbes typically found in the salt used to salt fish also play a role in degrading the fish. Water in an accessible form is necessary for the growth and metabolism of microorganisms. Halophiles cannot grow in media without salt; they thrive well in high salt concentrations. Although they can grow in quantities higher than that of seawater, halotolerant organisms thrive best in environments devoid of high salt content. While organisms classified as xerophiles develop rapidly in moderately dry environments or below an A_w of 0.85, osmophiles can flourish under high osmotic pressure (El Sheikha et al., 2014).

LAB and yeast are the primary microorganisms found in many fermented fish products (Mokoena et al., 2016). The main function of LAB is to reduce pH by fermenting available carbohydrates (El Sheikha et al., 2014). Some of the common LAB genera are *Weissella*, *Pediococcus*, *Tetragenococcus*, *Leuconostoc*, *Streptococcus* and *Lactobacillus* (Xu et al., 2021). Yeast, predominantly *Saccharomyces cerevisiae*, is another frequently found bacterium in fermented fish (Jespersen, 2003). Xu et al. (2021) reported that fermented fish products can include other yeast species such as *Candida tropicalis*, *Kluyveromyces marxianus*, *Zygosaccharomyces rouxii*, *Hansenula anomala*, *Candida zeylanoides* and *Rhodotorula glutinis* in addition to *S. cerevisiae*.

Many researchers have documented an extensive array of microbes implicated in the fermentation of fish in Africa. For example, the fermentation of *lanhouin* is linked to a wide variety of microbes (Anihouvi et al., 2006, 2007). Both Gram-positive and Gram-negative bacteria make up the microbial population of *lanhouin*. *Bacillus subtilis*, *Bacillus cereus*, *Bacillus licheniformis*, *Bacillus megaterium* and *Bacillus mycoides* constitute the majority of *Bacillus* isolates, while *Staphylococcus* species comprise *S. xylosus* and *Staphylococcus lentus*. Much like *lanhouin*, *momoni* is also fermented by various microorganism species, such as *Aspergillus*, *Bacillus*, *Lactobacillus*, *Pseudomonas*, *Pediococcus*, *Staphylococcus*, *Klebsiella*, *Debaryomyces* and *Hansenula* (Essuman, 1992; Nerquaye-Tetteh et al., 1978; Oronsaye, 1991; Sanni et al., 2002; Yankah, 1988). On the other hand, research by Koffi-Nevry et al. (2011) on *adjuevan*, a fermented fish from Côte d'Ivoire, reveals that LABs predominate in the fermentation process. The isolated and identified genera and species are *Lactobacillus fermentum*, *Leuconostoc lactis*, *Streptococcus* sp. and *Pediococcus* sp. According to Abd-Allah (2011), the microbiological load of fermented salted *M. cephalus* fish (*feseekh*) from Egypt includes *Staphylococcus equorum*, *B. subtilis*, *Lactobacillus* sp., and *Tetragenococcus halophilus*.

Microorganisms and enzymes both impact the flavour and texture of fermented fish products. For example, *S. cerevisiae* 31, *Lactobacillus plantarum* 120 and *S. xylosus* 135 are frequently used as starter cultures in fish fermentation, leading to the formation of desirable flavours due to their ability to produce esterase activities (Behera et al., 2018). The starting microbial culture used in fermented fish products affects some of its qualitative characteristics. Fermented fish mince, for instance, may gel in the presence of organic acids such as acetic or lactic acids because the pH lowering induces sufficient charge shifts and protein structural changes (Banwo et al., 2020). During the fish fermentation process, gel networks are created through disulphide bonds, non-disulphide covalent connections and hydrophobic contact. Even at the last fermentation phase, the gel network is further strengthened by the numerous disulphide connections (Xu et al., 2012). It is crucial to note that the gel qualities of fermented fish products are caused by both the activity of microbial and endogenous enzymes in the fermented fish products and the acidic environment produced by microbial fermentation (Xu et al., 2021).

6.6 Conclusion

Fermented meat and fish products hold a significant place in African heritage, ranging from *feseekh* and *khili* (North Africa) to *lanhouin and afo-nnama* (West Africa), *biltong* and *oshifima* (Southern Africa) and *mukeke* (East Africa), as testament to the ingenious approaches to meat and fish fermentation on the continent. The activity of beneficial microorganisms is central to the production of these traditional products, particularly LAB, yeasts, *Staphylococcus* and *Micrococcus* species, which enhance safety, shelf life and sensory qualities through biochemical and enzymatic transformations. These microorganisms start and control the fermentation process in meat and fish to produce and convert antimicrobial chemicals which aid in acidity, flavour development and preservation. Although the fermentation process naturally inhibits the growth of harmful microbes, incorrect handling or contamination may still be hazardous to consumers' health. Therefore, there is a need for ongoing research and education to ensure the microbiological quality of these products and to promote safe fermentation techniques.

References

Abd-Allah, S. (2011). Bacterial-flora of Egyptian salted *Mugil cephalus* fish (fessiekh) PCR-identification. *Assiut Veterinary Medical Journal*, *57*(130), 1–22.

Adeyeye, S. A. O., & Oyewole, O. B. (2016). An overview of traditional fish smoking in Africa. *Journal of Culinary Science & Technology*, *14*(3), 198–215. https://doi.org/10.1080/15428052.2015.1102785

Anihouvi, V., Ayernor, G., Hounhouigan, J., & Sakyi-Dawson, E. (2006). Quality characteristics of Lanhouin: A traditional processed fermented fish product in the Republic of Benin. *African Journal of Food, Agriculture, Nutrition and Development*, *6*(1), 1–15.

Anihouvi, V., Kindossi, J., & Hounhouigan, J. (2012). Processing and quality characteristics of some major fermented fish products from Africa: A critical review. *International Research Journal of Biological Sciences*, *1*(7), 72–84.

Anihouvi, V. B., Sakyi-Dawson, E., Ayernor, G. S., & Hounhouigan, J. D. (2007). Microbiological changes in naturally fermented cassava fish (*Pseudotolithus* sp.) for lanhouin production. *International Journal of Food Microbiology*, *116*(2), 287–291. doi:10.1016/j.ijfoodmicro.2006.12.009

Ashraf, S. A., Adnan, M., Patel, M., Siddiqui, A. J., Sachidanandan, M., Snoussi, M., & Hadi, S. (2020). Fish-based bioactives as potent nutraceuticals: Exploring the therapeutic perspective of sustainable food from the sea. *Mar Drugs*, *18*(5). doi:10.3390/md18050265

Bamidele, O. P., Adeyanju, A. A., Wokadala, O. C., & Mlambo, V. (2023). African Fermented fish and meat-based products. In *Indigenous fermented foods for the tropics* (pp. 117–131). Academic Press.

Baniga, Z., Dalsgaard, A., Mhongole, O. J., Madsen, H., & Mdegela, R. H. (2017). Microbial quality and safety of fresh and dried *Rastrineobola argentea* from Lake Victoria, Tanzania. *Food Control*, *81*, 16–22. doi:10.1016/j.foodcont.2017.05.023

Banwo, K., Ogunremi, O. R., & Sanni, A. I. (2020). Fermentation biotechnology of African traditional foods. In *Functional foods and biotechnology* (pp. 101–134). CRC Press.

Behera, S. S., Ray, R. C., & Zdolec, N. (2018). *Lactobacillus plantarum* with functional properties: An approach to increase safety and shelf-life of fermented foods. *BioMed Research International, 2018*, 9361614. doi:10.1155/2018/9361614

Belleggia, L., & Osimani, A. (2023). Fermented fish and fermented fish-based products, an ever-growing source of microbial diversity: A literature review. *Food Research International, 172*, 113112. doi:10.1016/j.foodres.2023.113112

Benkerroum, N. (2013). Traditional fermented foods of North African countries: Technology and food safety challenges with regard to microbiological risks. *Comprehensive Reviews in Food Science and Food Safety, 12*(1), 54–89.

Burfoot, D., Everis, L., Mulvey, L., Wood, A., & Campden, R. (2010). Literature review on microbiological hazards associated with biltong and similar dried meat products. *Food Standard Agency, London, 87.*

Carballo, J. (2012). The role of fermentation reactions in the generation of flavor and aroma of foods. *Fermentation: Effects on food properties* (pp. 51–83). CRC Press

Costello, C., Cao, L., Gelcich, S., Cisneros-Mata, M. A., Free, C. M., Froehlich, H. E., …, & Lubchenco, J. (2020). The future of food from the sea. *Nature, 588*(7836), 95–100. doi:10.1038/s41586-020-2616-y

Cruxen, C., Funck, G. D., Haubert, L., Dannenberg, G. D. S., Marques, J. L., Chaves, F. C., da Silva, W. P., & Fiorentini, A. M. (2019). Selection of native bacterial starter culture in the production of fermented meat sausages: Application potential, safety aspects, and emerging technologies. *Food Research International, 122*, 371–382. doi:10.1016/j.foodres.2019.04.018

Dirar, H. A. (1994). *Commentary: The fermented foods of the Sudan.*

El-Sebaiy, L. A., & Metwalli, S. M. (1989). Changes in some chemical characteristics and lipid composition of salted fermented Bouri fish muscle (*Mugil cephalus*). *Food Chemistry, 31*(1), 41–50.

El Sheikha, A., Montet, D., Panda, S., Worawattanamateekul, W., & Ray, R. (2014). *African fermented fish products in scope of risks. International Food Research Journal, 21*(2), 425–432.

Ercoskun, H., Tagi, S., & Ertas, A. H. (2010). The effect of different fermentation intervals on the quality characteristics of heat-treated and traditional sucuks. *Meat Science, 85*(1), 174–181. doi:10.1016/j.meatsci.2009.12.022

Essuman, K. M. (1992). *Fermented fish in Africa: A study on processing, marketing, and consumption*. Food & Agriculture Org.

Fraqueza, M. J., & Patarata, L. (2019). Fermented meat products: From the technology to the quality control. *Fermented food products* (pp. 197–238). CRC Press.

Gagaoua, M., & Boudechicha, H.-R. (2018). Ethnic meat products of the North African and Mediterranean countries: An overview. *Journal of Ethnic Foods, 5*(2), 83–98. doi:10.1016/j.jef.2018.02.004

Ghosh, S., Bornman, C., Meskini, M., & Joghataei, M. (2023). Microbial diversity in African foods and beverages: A systematic assessment. *Current Microbiology, 81*(1), 19. doi:10.1007/s00284-023-03481-z

Jespersen, L. (2003). Occurrence and taxonomic characteristics of strains of *Saccharomyces cerevisiae* predominant in African indigenous fermented foods and beverages. *FEMS Yeast Research*, *3*(2), 191–200. doi:10.1016/S1567-1356(02)00185-X

Joardder, M. U., & Masud, M. H. (2019). *Food preservation in developing countries: Challenges and solutions*. Springer.

Jones, M., Arnaud, E., Gouws, P., & Hoffman, L. C. (2017). Processing of South African biltong – A review. *South African Journal of Animal Science*, *47*(6). doi:10.4314/sajas.v47i6.2

Kandasamy, S., Kavitake, D., & Shetty, P. H. (2018). Lactic acid bacteria and yeasts as starter cultures for fermented foods and their role in commercialization of fermented foods. *Innovations in technologies for fermented food and beverage industries* (pp. 25–52). Springer.

Koffi-Nevry, R., Thibaut, O., Koussemon, M., & Brou, K. (2011). Chemical composition and lactic microflora of Adjuevan, A Traditional Ivorian Fermented Fish Condiment. *Pakistan Journal of Nutrition*, *10*. doi:10.3923/pjn.2011.332.337

Kouakou, A., N'guessan, f, Dadie, A., dje koffi, M., & Montet, D. (2012). Application of culture dependent methods and culture-independent methods (DGGE analysis) to study Lactic acid bacteria ecology of Ivorian fermented fish Adjuevan. *Challenges of Modern Technology*, *3*, 51–56.

Kouassi, A. T., Blei, S. H., Boko, A. C. E., & Angaman, D. M. (2024). Traditional fermented foods from Côte d'Ivoire: Knowledge, consumption and production methods in five localities surveyed. *International Journal of Gastronomy and Food Science*, *37*, 101002.

Koza, L., Drosinos, E., Cocolin, L., & Gasparik-Reichardt, J. (2008). Investigation of microbial association of traditionally fermented sausages. *Food Technology and Biotechnology*, *46*.

Kumar, P., Chatli, M. K., Verma, A. K., Mehta, N., Malav, O. P., Kumar, D., & Sharma, N. (2017). Quality, functionality, and shelf life of fermented meat and meat products: A review. *Critical Reviews in Food Science and Nutrition*, *57*(13), 2844–2856. doi:10.1080/10408398.2015.1074533

Laranjo, M., Potes, M. E., & Elias, M. (2019). Role of starter cultures on the safety of fermented meat products. *Frontiers in Microbiology*, *10*, 853. doi:10.3389/fmicb.2019.00853

Leroy, F., & De Vuyst, L. (2016). Fermented foods: Fermented meat products. In *Encyclopedia of food and health* (pp. 656–660). Elsevier.

Leroy, F., Verluyten, J., & De Vuyst, L. (2006). Functional meat starter cultures for improved sausage fermentation. *International Journal of Food Microbiology*, *106*(3), 270–285. doi:10.1016/j.ijfoodmicro.2005.06.027

Lücke, F.-K. (2017). Fermented meat products-an overview. *Fermented meat products-health aspects* (pp. 1–14). CRC Press.

Marijani, E., Charo-Karisa, H., Kigadye, E., & Okoth, S. (2020). Occurrence and exposure assessment of aflatoxin B1 in Omena (*Rastrineobola argentea*) from Kenya. *Journal of Food Quality*, *2020*(1), 8823340.

Mokoena, M. P., Mutanda, T., & Olaniran, A. O. (2016). Perspectives on the probiotic potential of lactic acid bacteria from African traditional fermented foods and beverages. *Food & Nutrition Research*, *60*(1), 29630.

Nahar, M. K., Zakaria, Z., Hashim, U., & Bari, M. F. (2017). Effect of pH and salt concentration on protein solubility of slaughtered and non-slaughtered broiler chicken meat. *Sains Malaysiana*, *46*(5), 719–724. doi:10.17576/jsm-2017-4605-06

Naidoo, K., & Lindsay, D. (2010). Survival of *Listeria monocytogenes*, and enterotoxin-producing *Staphylococcus aureus* and *Staphylococcus pasteuri*, during two types of biltong-manufacturing processes. *Food Control*, *21*(7), 1042–1050. doi:10.1016/j.foodcont.2009.12.025

Nerquaye-Tetteh, G., Eyeson, K., & Tete-Marmon, J. (1978). Studies on Bomone, a Ghanaian fermented fish product. *Ghana Journal of Agricultural Science*, *11*(1).

Oetterer, M., Perujo, S. D., Gallo, C. R., Arruda, L. F., Borghesi, R., & Cruz, A. M. P. (2003). Monitoring the sardine (*Sardinella brasiliensis*) fermentation process to obtain anchovies. *Scientia Agricola*, *60*, 511–517.

Oronsaye, J. (1991). An approach to fish processing and preservation. *Africa bio-science network (ABN)* (p. 125). Africa Bio-Science Network.

Petit, T., Caro, Y., Petit, A. S., Santchurn, S. J., & Collignan, A. (2014). Physicochemical and microbiological characteristics of biltong, a traditional salted dried meat of South Africa. *Meat Science*, *96*(3), 1313–1317. doi:10.1016/j.meatsci.2013.11.003

Petrovic, T. Z., Tomovic, V. M., Markovic, K. G., Semedo-Lemsaddek, T., & Grujovic, M. Z. (2025). Probiotics and honey: Boosting functional properties in dry fermented sausages. *Microorganisms*, *13*(2). doi:10.3390/microorganisms13020349

Sanni, A. I., Asiedu, M., & Ayernor, G. (2002). Microflora and chemical composition of momoni, a Ghanaian fermented fish condiment. *Journal of Food Composition and Analysis*, *15*(5), 577–583.

Strydom, P. E., & Zondagh, B. (2014). Biltong: A major South African ethnic meat product. In *Encyclopedia of meat sciences* (pp. 515–517). Elsevier.

Tamang, J. P., Cotter, P. D., Endo, A., Han, N. S., Kort, R., Liu, S. Q., Mayo, B., Westerik, N., & Hutkins, R. (2020). Fermented foods in a global age: East meets West. *Comprehensive Reviews in Food Science and Food Safety*, *19*(1), 184–217. doi:10.1111/1541-4337.12520

Todorov, S. D., Koep, K. S., Van Reenen, C. A., Hoffman, L. C., Slinde, E., & Dicks, L. M. (2007). Production of salami from beef, horse, mutton, Blesbok (*Damaliscus dorcas* phillipsi) and Springbok (*Antidorcas marsupialis*) with bacteriocinogenic strains of *Lactobacillus plantarum* and *Lactobacillus curvatus*. *Meat Science*, *77*(3), 405–412. doi:10.1016/j.meatsci.2007.04.007

Van der Riet, W. (1982). Biltong-A south-African dried meat product. *Fleischwirtschaft*, *62*(8), 970-&.

Xiao, Y., Li, P., Zhou, Y., Ma, F., & Chen, C. (2018). Effect of inoculating *Lactobacillus pentosus* R3 on N-nitrosamines and bacterial communities in dry fermented sausages. *Food Control*, *87*, 126–134. doi:10.1016/j.foodcont.2017.12.025

Xu, Q., Li, S., Huang, H., & Wen, J. (2012). Key technologies for the industrial production of fumaric acid by fermentation. *Biotechnology Advances*, *30*(6), 1685–1696. doi:10.1016/j.biotechadv.2012.08.007

Xu, Y., Zang, J., Regenstein, J. M., & Xia, W. (2021). Technological roles of microorganisms in fish fermentation: a review. *Critical Reviews in Food Science and Nutrition*, *61*(6), 1000–1012. doi:10.1080/10408398.2020.1750342

Yankah, V. (1988). Studies on momone: A Ghanaian fermented fish product. *B. Sciences thesis*. Department of Nutrition and Food Science, University of Ghana–Legon.

Zakpaa, H., Mak-Mensah, E., & Johnson, F. (2009). Production of bio-ethanol from corncobs using *Aspergillus niger* and *Saccharomyces cerevisiae* in simultaneous saccharification and fermentation. *African Journal of Biotechnology*, *8*, 3018–3022.

Zang, J., Xu, Y., Xia, W., & Regenstein, J. M. (2020). Quality, functionality, and microbiology of fermented fish: a review. *Critical Reviews in Food Science and Nutrition*, *60*(7), 1228–1242. doi:10.1080/10408398.2019.1565491

7

Microbiology of African Fermented Vegetable and Fruit-Based Products

7.1 Introduction

The World Health Organization (WHO) and Food and Agriculture Organization (FAO) advocate consuming vegetables and fruits every day to help prevent chronic conditions such as stroke, coronary heart disease and high blood pressure. Fresh, nutrient-dense, health-promoting and readily consumable foods and beverages are generally preferred by consumers (Demir et al., 2006). Vegetables and fruits are frequently fermented to preserve and enhance their nutritional and sensory value (Di Cagno et al., 2013; Swain et al., 2014). In Africa, fermentation is still mostly home-based, and different raw materials such as fruits, vegetables, cereals and legumes are used (Olasupo et al., 2010). Most vegetables and fruits are either eaten raw or processed industrially to make pastes, drinks, sauces, soups, salad dressings and canned and dried juices. Fresh produce, especially when minimally treated, has a short shelf life due to its susceptibility to pathogen contamination and rapid microbial spoilage. The primary preservation technologies include pasteurising, cooking and adding chemical preservatives to ensure the safety of vegetables and fruits, which often result in unfavourable modifications to their chemical composition and physical characteristics (Di Cagno et al., 2013; Saud et al., 2024). To mitigate these drawbacks, fermentation is regarded as an effective and straightforward technique that can either preserve and/or improve nutritional value, sensory appeal and the shelf life of vegetables and fruits (Saud et al., 2024; Yuan et al., 2024).

DOI: 10.1201/9781003658320-7

Vegetables and fruits undergo fermentation through a variety of microbes, and the microbiome is highly diverse. However, the microbes that are naturally present in these products will ultimately cause them to spoil. In general, spoiling organisms are more prevalent and proliferate faster than fermenting microorganisms. The fermentation process of vegetable and fruit-based products predominantly relies on lactic acid bacteria (LAB), particularly the *Weissella*, *Leuconostoc*, *Pediococcus* and *Tetragenococcus* species (Chang et al., 2008; Hutkins, 2008; Watanabe et al., 2009). These organisms produce distinctive byproducts, including lactic and acetic acids, ethyl alcohol and CO_2 (Montville et al., 2012).

Fermentation involving vegetables and fruits falls into one of four categories: lactic acid, alcoholic, alkaline or acetic acid (Nout, 2014; Taale et al., 2024). Although this technique is less known, alkaline-fermented vegetables are primarily consumed in Asian and African countries (Ayua & Omware, 2013; Nout, 2014; Vouidibio Mbozo et al., 2017). However, in general, the fermentation of vegetables and fruits are not well practised on the African continent. Fruits and leafy vegetables that have undergone fermentation may benefit in terms of improved nutrition, longer shelf life, increased safety and easier access to certain vitamins, trace minerals and antioxidants. This chapter discusses the microbiology of the fermentation of developing and classic African vegetables and fruit products and their main characteristics.

7.2 Different Fermented African Vegetables and Fruits

On the African continent, vegetables are fermented to produce sauces, soups and condiments, while fruits are primarily processed to formulate alcoholic and non-alcoholic beverages (Mandha et al., 2021). Several value-added fermented vegetable and fruit-based products, including smoothies made from a blend of Chayote leaf and pineapple, purées made from *papaya*, juices and wine, have also been developed (Managa et al., 2021; Mashitoa et al., 2021; Minnaar et al., 2017). Other vegetables and fruits, such as *makalani* palm, bird plum, *marula*, jackal berry, *manketti*, *jujube*, large false *mopane*, wild loquat, buffalo thorn and sand apple, are frequently fermented to develop various products in Africa (Misihairabgwi & Cheikhyoussef, 2017).

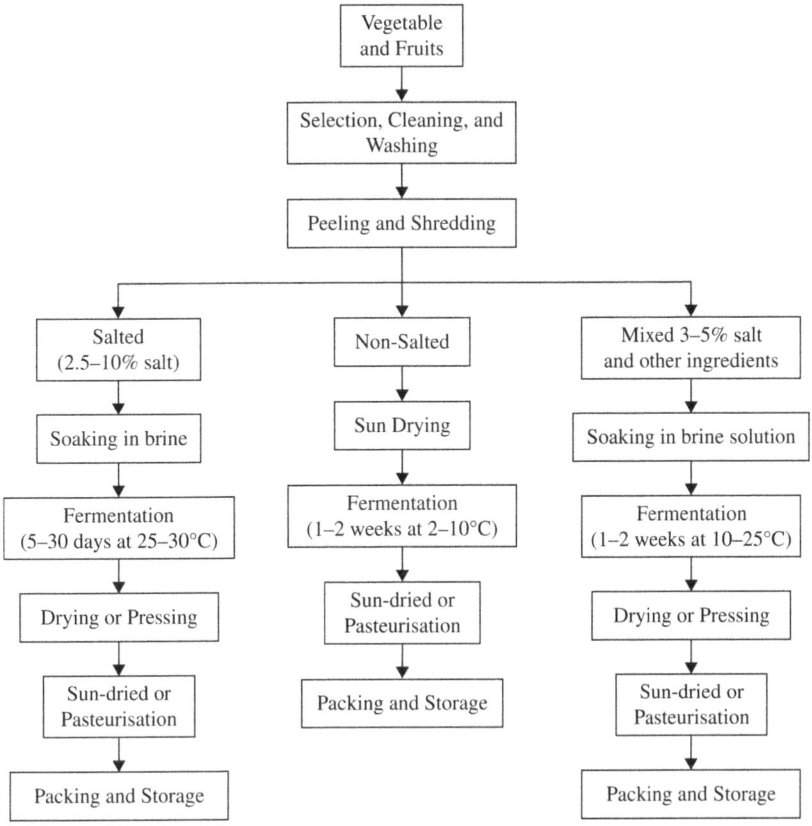

Figure 7.1 The basic fermentation process of vegetables and fruits. (Adapted from Swain et al., 2014.)

This chapter presents notable fermented vegetable and fruit products from African regions, and Figure 7.1 provides a general overview of their fermentation.

7.2.1 Selected Fermented Vegetables and Fruit Products from Central Africa

7.2.1.1 Ntoba mbodi

Ntoba mbodi (Figure 7.2) is a common leafy vegetable condiment in the Democratic Republic of Congo made from fermented cassava (*Manihot esculenta*). Cassava leaves are gathered and wilted for 23 days. After being cleaned and cut into smaller pieces, the wilted leaves are washed in water. The washed leaves are divided into small pieces and wrapped in large green leaves, typically from fresh *Carica papaya* or *Cyrtosperma senegalense* plants. After fermentation for 24 days in ambient conditions,

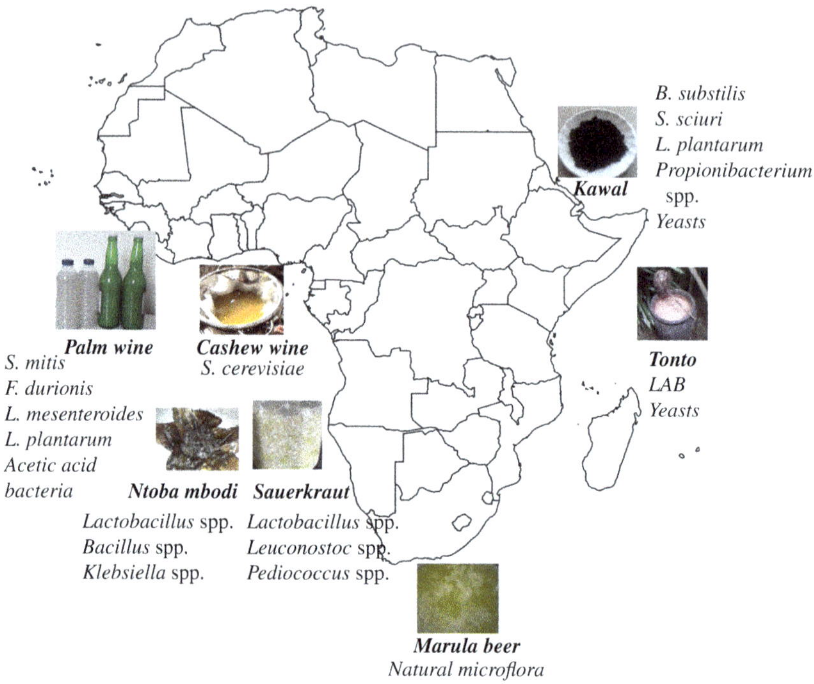

Figure 7.2 A pictorial view of various African vegetable and fruit products from different regions of Africa and the distinct fermenting microorganisms. (Palm wine, *tonto* (From Gbashi et al., 2023), *kawal* (From Lawane et al., 2020), *ntoba mbodi* (From NzaouAimé & Kayath, 2016), cashew wine (https://friendsofbib.com/cashew-wine-juice-and-nuts/), *marula* beer (https://bottelenbere.weebly.com/marula-beer.html), *saukeraut* (https://www.whatscookingella.com/homemade-sauerkraut.html).)

the product is now considered ready for consumption. Cassava leaf fermentation is essential because it reduces or removes the toxic substances inside the leaves (Ukaegbu et al., 2019).

7.2.1.2 *Urugwa*

In Rwanda, *urugwa* is a beer made with bananas. Ripe banana fruits are peeled and crushed to extract the juice. Clean water is poured over the juice and then roasted sorghum grain is added. Usually, the mixture is left in a pit enclosed with banana leaves to ferment for 24 days. The beer has a short shelf life as neither the raw ingredients nor the banana beer is pasteurised or sterilised. The finished product should be kept in cold storage to avert the proliferation of spoilage and pathogenic organisms. Therefore, it is important to use clean equipment and follow hygienic methods when making this beer (Oguntoyinbo et al., 2016).

7.2.1.3 Sauerkraut

Although fermented cabbage is more frequently associated with European cuisine, countries such as Cameroon and Gabon in Central Africa produce varieties of *sauerkraut*. *Sauerkraut* (Figure 7.2) is made through a relatively easy technique called spontaneous fermentation. The first step in preparing *sauerkraut* is to cut fresh cabbage using either a sharp knife, slicer or food processor into thin strips, after which it is transferred to a large mixing bowl and sprinkled with approximately 1.5–2% of the weight of the cabbage in salt. Thereafter, the shredded cabbage is manually massaged with the salt to break down its cell walls and release its natural juices, fostering the fermentation process. The cabbage is placed in a sterile vessel such as a food-grade plastic container, glass jar or ceramic crock to let it wilt and release sufficient liquid. The container is stored in a cold, dark area, preferably at room temperature (approximately 65–75°F/18–24°C) to enable fermentation, which often involves *Leuconostoc*, *Lactobacillus* and *Pediococcus* species for one to four weeks (Dimidi et al., 2019). When the *sauerkraut* reaches the proper sourness, it is ready to eat and should be transferred into sterile, sealed jars and refrigerated to slow down the fermentation process. *Sauerkraut* stored under the right conditions can keep for several months or even longer.

7.2.2 Selected Fermented Vegetables and Fruit Products from West Africa

7.2.2.1 Agadadigi

Agadagidi is a popular beverage in Western Nigeria, consumed at ceremonies. The product is made from ripe plantains obtained from markets. The fruit is then preserved for 45 days until it is very soft and overripe. After washing to remove dirt, the soft fruits are peeled, chopped into small pieces, submerged in water (1:5 w/v), covered and allowed to incubate for approximately three days at room temperature. After fermentation, the mixture is filtered to extract the liquid from plantain pulp, making it accessible to customers as an alcoholic beverage (Folakemi et al., 2012).

7.2.2.2 Cashew Wine

Cashew is a significant nut crop with a unique flavour used to make cashew wine in Nigeria (Awe & Olayinka, 2011). To make cashew wine (Figure 7.2), ripe cashew apples are gathered and cleaned. After washing, the apples are chopped into small pieces and mixed and warm water (45°C) is added to dilute the pulp. After the addition of 30% sucrose, a standardised Campden tablet (a sulphur-based product used for sterilisation) and yeast nutrients, the mixture ferments for 24 hours (at 25°C). The fermentation is stopped after six days, and the fermentum is sieved to remove the fibrous and solid parts (Awe & Olayinka, 2011). The product is yellow-coloured, with an alcohol concentration of 6–12% (v/v).

7.2.2.3 Palm Wine

In several West African countries, palm wine (Figure 7.2) is a common fermented alcoholic beverage. This is made from the sap of *Borassus akeassii*, *Elaeis guineensis*, *Raphia hookeri*, *Raphia guineensis* and *Borassus aethiopum*. In some parts of Nigeria, where it is known as *nkwu*, *ngwommanya* or *ocha*, palm wine has significant cultural importance. Large servings of the beverage are provided to attendees during traditional ceremonies such as weddings, funerals and birthday celebrations. Sometimes, a mixture of palm wine and medicinal herbs is used to treat various illnesses. The carbohydrate compounds present, including glucose, sucrose, maltose and raffinose, are fermented by *Saccharomyces cerevisiae* to make alcohol. Apart from yeasts, researchers have observed that bacteria, including *Leuconostoc, Lactobacillus* and acetic acid bacteria, aid in the fermentation of palm wine (Nwaiwu & Itumoh, 2017).

7.2.3 Selected Fermented Vegetables and Fruit Products from East Africa

7.2.3.1 Tonto

Tonto or *mwenge bigere* is a fermented banana beer commonly consumed in Uganda (Figure 7.1). To make banana beer, banana pulp juice is fermented with a cereal flour. The beer is a little cloudy but has a pleasant flavour and keeps for a few days (Thierry & Osinachi, 2023). Ripe bananas are typically utilised due to their high sugar content. The production process includes extracting and diluting banana juice and milling cereals into flour. The milled cereal (such as sorghum or millet) is combined with diluted banana juice and allowed to ferment for 18–24 hours (Padonou et al., 2023). Yeasts are the naturally occurring microbes that ferment the banana fruit (Oguntoyinbo et al., 2016). Following fermentation, the banana beer (alcohol percentage of 6–11% v/v) is filtered through a cotton cloth before being offered to customers.

7.2.3.2 *Kawal*

Kawal is used in soups and sauces as a meat alternative (Figure 7.1). It is made from the leaves of the wild legume *Cassia obtusifolia* L., which grows in Sudan and Chad (Dirar et al., 2006; Mbaiguinam et al., 2005). These leaves are widely available and are rich in oxalates, proteins and amino acids. The process of making *kawal* entails gathering the leaves and removing the unwashed stalks, pods and flowers. The leaves should not be washed because the natural microbes on them are crucial to the fermentation of the final product. Using a mortar and pestle, the leaves are ground into a paste without any juice being extracted. Thereafter, this paste is put into an earthen jar and covered with plant leaves. The jar is then buried up to the neck in the earth after being coated with mud. The contents are manually combined every three days, and fermentation takes approximately 14 days to complete (Mbaiguinam et al., 2005). The fermentation process is extremely complicated, requiring LAB, such as *Lactiplantibacillus plantarum*, yeasts, *Propionibacterium*, *Staphylococcus sciuri*, and *Bacillus subtilis* (Dirar et al., 2006). After fermentation, a pungent black paste is formed into balls and left to sun-dry for approximately 35 days.

7.2.4 *Selected Fermented Vegetables and Fruit Products from Southern Africa*

7.2.4.1 *Masau*

The fruits of the buffalo thorn, known as *masau* in Zimbabwe, are usually eaten either raw or processed into products such as *mahewu* (a fermented beverage), porridge, jam and traditional cakes. In addition, the fruits are fermented with maize to produce the alcoholic beverage *kachasu*. Traditionally, *masau* is fermented spontaneously to produce beverages. Other African wild fruits that spontaneously ferment are *mazhanje*, *marula* and *hacha* (Nyanga et al., 2007). However, the microbiota and quality of the fermented products are inconsistent due to the uncontrolled or spontaneous fermentation of these fruits. Their fermentation mostly depends on the surface microbes on the fruit and the tools employed in the process.

7.2.4.2 Mudetemwa

In Zimbabwe, the fruit of the *Parinari curatellifolia* plant (sand apple) is used to make *mudetemwa*. The juice is extracted from the pounded fruits through boiling, which is allowed to ferment overnight and then boiled again. After boiling, the beer is cooled before being served to customers. The sugar plum tree (*Uapaca kirkiana*) fruit is another fruit valued in Zimbabwe. This fruit is fermented to produce *mutandavira*, a type of alcoholic wine. The fruit of the sugar plum tree is first mashed to remove the skin and seeds from the pulp to make *mutandavira*. After that, the extracted pulp is combined with cold water and allowed to ferment to produce the wine (Oguntoyinbo et al., 2016).

7.2.4.3 Marula Beer

Marula beer (Figure 7.1) is commercially produced in South Africa and is available internationally under the *Amarula* liquor brand. Locally, the first step in making *marula* beer is harvesting the ripe fruit from the *marula* tree. The tree is an African botanical gem that has been used for hundreds of years to provide people with nutritional benefits such as calcium, magnesium, vitamin C and potassium. After the fruit is gathered, cleaned and peeled, it is pressed, mashed or smashed to extract the juice. The appropriate amount of water is added during this procedure to give the juice a soup-like as opposed to a paste-like consistency. The clear juice settles to the bottom of the jar while the fuzzy foam floats to the top after the juice has been covered and allowed to ferment for a few hours. Depending on the desired alcohol strength, the clear juice is allowed to ferment for 24 days after the top layer of foam has been removed. On a very limited scale, a variety of native African fruits, such as coconut, grapefruit, mango, plum, apple, *marula* and cider, are also fermented into vinegar (Ameyapoh et al., 2010; Ezemba et al., 2021; Oguntoyinbo et al., 2016). The fruit of the so-called sausage tree is used to make the alcoholic beverage known as *muratina*. Although indigenous to South Africa, the tree grows throughout Kenya and Northern Africa (Harkishor, 1977; Oguntoyinbo et al., 2016).

7.3 Microorganisms Involved in the Fermentation of African Vegetable and Fruit Products

The microbial transformations that occur in fermented foods are primarily dominated by catalysis and the conversion of compounds or substances into newer and smaller ones. These transformations are driven by enzymes produced by the microbes and can involve reactions such as oxidation, reduction and hydrolysis. Organic compounds, such as amino acids, carbohydrates, short-chain organic acids, polyphenols and bioactive peptides, may be metabolised during the fermentation process (Jimoh & Adebo, 2025). The type of microorganisms that predominate in the fermentative media often determines the microbial transformation pathways. Apart from the catabolic processes microorganisms perform, other substances, such as acids, esters, polyphenols, alcohols, bioactive peptides and vitamins, may also be synthesised (Jimoh & Adebo, 2025; Shah & Singhal, 2017). Consequently, the catabolic and anabolic processes impact the sensory qualities, nutritional value and microbiological safety of the food (Septembre-Malaterre et al., 2018). Various types of fermentation can occur in vegetables and fruits, namely, alcoholic, propionic, lactic, malolactic and butyric fermentation (Septembre-Malaterre et al., 2018).

Vegetables and fruits can be fermented through natural fermentation (spontaneous fermentation) and/or by adding starter cultures (controlled fermentation) (Table 7.1). Natural surface microflora, such as *Lactobacillus* spp., *Leuconostoc* spp. and *Pediococcus* spp., can ferment vegetables and fruits spontaneously; however, using starter cultures, such as *Lactobacillus acidophilus*, *Lactobacillus gasseri*, *Lactobacillus rhamnosus* and *L. plantarum*, provides consistency and reliability of performance (Swain et al., 2014). Numerous investigations have detailed the diverse biochemical interactions that occur during the natural fruit and vegetable fermentation processes. For example, after the spontaneous fermentation of sicklepod, researchers attributed the reduction of biochemical contents to *L. plantarum*, *S. sciuri* and *B. subtilis* that metabolise carbohydrates during fermentation (Ali et al., 2020; Ibrahim et al., 2019; Osman et al., 2009). However, to maintain the fermentation process in some leafy vegetables, a source of carbohydrates should be added to the medium containing leafy vegetables (Kasangi et al., 2010).

The most researched lactic-acid-fermented vegetables are fermented cucumbers and cabbage (*sauerkraut*), which are made by immersing vegetables in brine (2–5% salt) (Behera et al., 2020; Swain et al., 2014). Therefore, by producing exopolysaccharides, bacteriocins and heterogeneous compounds, LAB help extend shelf life and improve the sensory qualities of fermented vegetable pickles (Behera et al., 2020). In addition

Table 7.1 Microorganisms Involved in the Fermentation of African Vegetables and Fruits

FERMENTED PRODUCT	MICROBES INVOLVED IN FERMENTATION	COUNTRY	REFERENCES
Palm wine *emu* or *oguro*	*Streptococcus mitis, Fructobacillus durionis, Leuconostoc. mesenteroides* and *Lactiplantibacillus plantarum*. Acetic acid bacteria. *Schizosaccharomyces, Trigonopsis, Galactomyces, Candida, Trichosporon, Hanseniaspora, Kodamaea, Arthroascus* and *Issatchenkia*	Africa	Ehrmann et al. (2009), Faparusi and Bassir (1972) and Ouoba et al. (2012)
Ntoba mbodi	*Lactobacillus* spp., *Bacillus* spp. and *Klebsiella* spp.	Congo	Ukaegbu et al. (2019)
Tella	Yeast and LAB	Ethiopia	Faparusi (1973)
Agadagidi	*Saccharomyces, Leuconostoc, Streptococcus, Bacillus* and *Micrococcus*	Nigeria	Sanni (1989)
Cashew wine	*Saccharomyces cerevisiae* SIL 59703	Nigeria	Awe and Olayinka (2011)
Cocoa wine	*S. cerevisiae*	West Africa	Ludlow et al. (2016)
Kombucha	*Acetobacter, Gluconobacter, Saccharomyces* spp. and non-*Saccharomyces* spp.	North Africa, Middle East	Malbasa et al. (2015) and Rahmani et al. (2019)
Muratina	*S. cerevisiae, Lacticaseibacillus casei*	Kenya	Dimidi et al. (2019) and Jespersen (2003)
Marula fruit juice	Spontaneous fermentation	South Africa	Hiwilepo-van Hal et al. (2013)
Plantain fruit	Yeast, aerotolerant propionic acid bacteria and *Lactobacillus*	West Africa	Atunnise et al. (2020)
African kale leaves	*Limosilactobacillus fermentum* BFE 6620 and *L. plantarum* BFE 5092	Africa	Odongo et al. (2017)

to the acidic LAB fermentation of vegetables, vegetable proteins such as the African locust bean, cottonseeds, melon, *baobab* seed, *bambara* groundnut, castor oil bean, African mesquite and roselle seed can undergo alkaline fermentation (Fusco et al., 2017). Microorganisms such as filamentous fungi and *Bacillus* species release enzymes that hydrolyse proteins into amino acids, peptides, polypeptides and ammonia (Nout, 2014; Oguntoyinbo et al., 2016). After 60 hours of the spontaneous fermentation of *baobab* seeds (a condiment from West Africa), the concentration of total free amino acids increased due to microbial metabolic activities (Parkouda et al., 2015). As the microbes hydrolyse the proteins in the seeds, the observed alterations were attributed to their proteolytic activity.

The two most common microbial groups that ferment vegetable proteins are *Bacillus* and *Staphylococcus* species (Ahaotu et al., 2013; Ouoba et al., 2019). Some microorganisms that cause vegetables to ferment exhibit antagonistic effects on pathogen growth and survival in the food. However, incorrect handling and using contaminated packaging material can result in microbiological risks even after processing (Adinsi et al., 2017; Mpofu et al., 2016; Schoder et al., 2013).

The literature provides numerous examples of the identification of microbes which play a role in African fermented vegetables and fruits. For example, *Bacillus macerans*, *B. subtilis*, *Micrococcus varians*, *S. sciuri* and *Staphylococcus xylosus* have been detected by phenotypically based characterisation derived from media and culture isolation (Louembe et al., 2003). Kobawila et al. (2005) reported the presence of *Bacillus amyloliquefaciens*, *Bacillus brevis*, *Bacillus polymixa*, *B. subtilis*, *B. macerans*, *Bacillus cereus*, *Bacillus pumilus* and *Bacillus megaterium* using the same microbial characterisation approach. Using both phenotypic and genotypic approaches, Vouidibio Mbozo et al. (2017) found *Bacillus safensis*, *Lysinibacillus* spp. and *Bacillus siamensis* in *ntoba mbodi*. The genotypic methods used for microbial detection include 16S-23S rDNA internal transcribed spacer (ITS), repetitive sequence-based (rep)-PCR and 16S rRNA gene sequencing. Ouoba et al. (2015) also discovered a new species of bacteria belonging to the genus *Lysinibacillus* in *ntoba mbodi*.

Fruit juices may have modified sensory profiles due to LAB activities during fermentation. Original qualities may have been altered due to the presence of amino acids, volatile chemicals (acids, ketones, aldehydes, hydrocarbons, alcohols and esters) and organic acids (malic, citric and lactic acids) (Rodríguez et al., 2021). According to Nyanga et al. (2007) and Wilson et al. (2012), fruit fermentation can include both lactic acid fermentation and alcoholic fermentation, particularly when making wines and African brews. LAB and yeasts have a symbiotic relationship in which the LAB creates an acidic environment that is conducive to yeast growth, and the yeasts subsequently produce vitamins and amino acids to support LAB growth (Wilson et al., 2012). An example of such relationships in play is fermented *jujube* fruit pulp and Rwandese banana (*urwagwa*) beer (Nyanga et al., 2007; Wilson et al., 2012).

The fermentation of bananas and plantain fruit to produce *agadagidi* alcohol involves microbial species, such as *Leuconostoc mesenteroides, L. plantarum* and *B. subtilis. Bacilli* are spore-forming, naturally occurring bacteria that can withstand severe environments (Alo et al., 2012). Therefore, the occurrence of *B. subtilis* during *agadagidi* production could be due to the presence of the bacterium's endospores from the peels, dust or the air, which then contaminate the fermentation substrate. The yeasts, *Candida utilis* and *S. cerevisiae,* are present at every stage of *agadagidi* production, which could be due to the composition and environment of the fruit. Fruit that has reached maturity typically includes yeasts, which play a key role in alcohol production. The moulds in the skin could potentially spread to the plantain pulp. Mould eradication most likely starts due to LAB multiplication and enhanced activity during fermentation. Other isolated bacteria are antagonistic to yeasts and LAB, suggesting that these microorganisms may create inhibitory compounds (Dalié et al., 2010).

7.4 Conclusion

The microbiology of African fermented vegetables and fruits is rich and diversified due to a complex community of microorganisms responsible for their distinctive aromas, textures and nutritional profiles. Prominent microbial species such as *Saccharomyces, Bacillus* spp. and LAB (e.g., *L. plantarum* and *L. mesenteroides*) are essential to the fermentation of African vegetables and fruits. By preventing harmful and spoilage organisms from growing, these microbes help to acidify the environment of the food and increase its shelf life. African fermented vegetables and fruits have a symbiotic interaction between human culture and microbiological life that not only supports significant culinary traditions but also presents opportunities for innovation in food science, nutrition and sustainable food systems.

References

Adinsi, L., Mestres, C., Akissoé, N., Vieira-Dalodé, G., Anihouvi, V., Durand, N., & Hounhouigan, D. J. (2017). Comprehensive quality and potential hazards of gowe, a malted and fermented cereal beverage from West Africa. A diagnostic for a future re-engineering. *Food Control*, *82*, 18–25. doi:10.1016/j.foodcont.2017.06.019

Ahaotu, I., Anyogu, A., Njoku, O. H., Odu, N. N., Sutherland, J. P., & Ouoba, L. I. (2013). Molecular identification and safety of *Bacillus* species involved in the fermentation of African oil beans (*Pentaclethra macrophylla Benth*) for production of Ugba. *International Journal of Food Microbiology*, *162*(1), 95–104. doi:10.1016/j.ijfoodmicro.2013.01.001

Ali, M. A. M., Musa, R., Ali, A. A., Ali, Y., & Hamadnalla, H. M. (2020). Isolation and identification of bacteria from dried fermented leaves of *Cassia obtusifolia*. L (Kawal). *Clinical Medicine and Medical Research*, *1*(1), 16–20.

Alo, M., Anyim, C., Igwe, J., Elom, M., & Uchenna, D. (2012). Antibacterial activity of water, ethanol and methanol extracts of *Ocimum gratissimum*, *Vernonia amygdalina* and *Aframomum melegueta*. *Advances in Applied Science Research*, *3*(2), 844–848.

Ameyapoh, Y., Leveau, J.-Y., Karou, S. D., Bouix, M., Sossou, S. K., & De Souza, C. (2010). Vinegar production from Togolese local variety Mangovi of mango *Mangifera indica Linn*. (*Anacardiaceae*). *Pakistan Journal of Biological Sciences: PJBS*, *13*(3), 132–137.

Atunnise, A. K., Salau, T. B., Odufuwa, K. T., Olukanni, D. O., Adeyanju, M. M, & Salau, B. A. (2020). Enhancement of Bioactive Compounds During Fermentation of Unripe Musa paradisiaca. *Trends in Applied Sciences Research*, *15*, 115–124.

Awe, S., & Olayinka, E. T. (2011). Effect of cashew wine on histology and enzyme activities of rat liver. *Journal of Toxicology and Environmental Health Sciences*, *3*(2), 31–38.

Ayua, E., & Omware, J. (2013). Assessment of processing methods and preservation of African leafy vegetables in Siaya county, Kenya. *Global Journal of Biology Agriculture and Health Sciences*, *2*(2), 46–48.

Behera, S. S., El Sheikha, A. F., Hammami, R., & Kumar, A. (2020). Traditionally fermented pickles: How the microbial diversity associated with their nutritional and health benefits? *Journal of Functional Foods*, *70*, 103971.

Chang, H. W., Kim, K. H., Nam, Y. D., Roh, S. W., Kim, M. S., Jeon, C. O., Oh, H. M., & Bae, J. W. (2008). Analysis of yeast and archaeal population dynamics in kimchi using denaturing gradient gel electrophoresis. *International Journal of Food Microbiology*, *126*(1–2), 159–166. doi:10.1016/j.ijfoodmicro.2008.05.013

Dalié, D. K. D., Deschamps, A. M., & Richard-Forget, F. (2010). Lactic acid bacteria – Potential for control of mould growth and mycotoxins: A review. *Food Control*, *21*(4), 370–380. doi:10.1016/j.foodcont.2009.07.011

Demir, N., Bahceci, K. S., & Acar, J. (2006). The effects of different initial *Lactobacillus plantarum* concentrations on some properties of fermented carrot juice. *Journal of Food Processing and Preservation*, *30*(3), 352–363. doi:10.1111/j.1745-4549.2006.00070.x

Di Cagno, R., Coda, R., De Angelis, M., & Gobbetti, M. (2013). Exploitation of vegetables and fruits through lactic acid fermentation. *Food microbiology*, *33*(1), 1–10. doi:10.1016/j.fm.2012.09.003

Dimidi, E., Cox, S. R., Rossi, M., & Whelan, K. (2019). Fermented foods: Definitions and characteristics, impact on the gut microbiota and effects on gastrointestinal health and disease. *Nutrients*, *11*(8). doi:10.3390/nu11081806

Dirar, H. A., Harper, D. B., & Collins, M. A. (2006). Biochemical and microbiological studies on kawal, a meat substitute derived by fermentation of *Cassia obtusifolia* leaves. *Journal of the Science of Food and Agriculture*, *36*(9), 881–892. doi:10.1002/jsfa.2740360919

Ehrmann, M. A., Freiding, S., & Vogel, R. F. (2009). *Leuconostoc palmae* sp. nov., a novel lactic acid bacterium isolated from palm wine. *International Journal of Systematic and Evolutionary Microbiology*, *59*(5), 943–947.

Ezemba, A., Osuala, E. O., Orji, M., Ezemba, C. C., & Anaukwu, C. G. (2021). Production and comparative physicochemical analysis of vinegar from locally grown fruits in Nigeria and industrial produced vinegar. *American Journal of Microbiological Research*, *9*(1), 25–33.

Faparusi, S. (1973). Origin of initial microflora of palm wine from oil palm trees (*Elaeis guineensis*). *Journal of Applied Microbiology*, *36*(4), 559–565.

Faparusi, S., & Bassir, O. (1972). Factors affecting palm wine. Period of tapping. *West African Journal of Biological and Applied Chemistry*, *15*, 17–23.

Folakemi, O. P., Flora, D. O., & Monday, K. R. (2012). The effect of chemical preservatives, pasteurization and refrigeration on the shelf life of agadagidi A fermented plantain drink. *Research Journal of Microbiology*, *7*(3), 145.

Fusco, V., Oguntoyinbo, F. A., & Franz, C. M. A. P. (2017). Fermentation to improve food security in Africa and Asia. In *Soft chemistry and food fermentation* (pp. 337–378). Academic Press.

Gbashi, S., Moyo, S. M., Olopade, B., Kewuyemi, Y., Areo, O. M., Lawal, O. M., Momoh, C. O., Igbashio, M. D., & Njobeh, P. B. (2023). African fermented vegetable and fruit-based products. In *Indigenous fermented foods for the tropics* (pp. 227–244). Academic Press.

Harkishor, K. (1977). Kenyan sugarcane wine-muratina. Paper presented at the Proceedings of the Symposium of Indigenous Fermented Food, Bangkok.

Hiwilepo-van Hal, P., Bille, P. G., Verkerk, R., & Dekker, M. (2013). The effect of temperature and time on the quality of naturally fermented marula (*Sclerocarya birrea* subsp. *Caffra*) juice. *LWT – Food Science and Technology*, *53*(1), 70–75. doi:10.1016/j.lwt.2013.02.021

Hutkins, R. W. (2008). *Microbiology and technology of fermented foods*. John Wiley & Sons.

Ibrahim, N. A., Ibrahim, N. A., Elgasim, E. A., & Mustafa, S. A. (2019). Isolation and identification of microorganisms in kawal (*Cassia obtusifolia*) fermented leaves. *International Journal of Current Research*, *11*(10), 7956–7960. doi:10.24941/ijcr.36741.10.2019

Jespersen, L. (2003). Occurrence and taxonomic characteristics of strains of *Saccharomyces cerevisiae* predominant in African indigenous fermented foods and beverages. *FEMS Yeast Research*, *3*(2), 191–200. doi:10.1016/S1567-1356(02)00185-X

Jimoh, A. A., & Adebo, O. A. (2025). Evaluation of antiobesogenic properties of fermented foods: In silico insights. *Journal of Food Science*, *90*(3), e70074. doi:10.1111/1750-3841.70074

Kasangi, D., Shitandi, A., Shalo, P., & Mbugua, S. (2010). Effect of spontaneous fermentation of cowpea leaves (*Vigna unguiculata*) on proximate composition, mineral content, chlorophyll content and beta-carotene content. *International Food Research Journal*, *17*(3), 721–732.

Kobawila, S. C., Louembe, D., Keleke, S., Hounhouigan, J., & Gamba, C. (2005). Reduction of the cyanide content during fermentation of cassava roots and leaves to produce bikedi and ntoba mbodi, two food products from Congo. *African Journal of Biotechnology*, *4*(7), 689–696. doi:10.5897/ajb2005.000-3128

Lawane, A. I., Tidjani, A., Parkouda, C., François, T., Hissein, A., & Aly, S. (2020). Amino acid profiling of the leaves of senna *Obtusifolia* and biochemical changes during fermentation for production of "Kawal", a traditional food condiment. *Food Processing & Nutritional Science*, *1*(2), 145153.

Louembe, D., Kobawila, S., Bouanga-Kalou, G., & Kéléké, S. (2003). Microbiological study of fermented cassava leaves: "*Ntoba Mbodi*". *Tropicultura*, *21*(3), 106–111.

Ludlow, C. L., Cromie, G. A., Garmendia-Torres, C., Sirr, A., Hays, M., Field, C., Jeffery, E. W., Fay, J. C., & Dudley, A. M. (2016). Independent origins of yeast associated with coffee and cacao fermentation. *Current Biology*, *26*(7), 965–971. doi:10.1016/j.cub.2016.02.012

Malbasa, R., Jevric, L., Loncar, E., Vitas, J., Podunavac-Kuzmanovic, S., Milanovic, S., & Kovacevic, S. (2015). Chemometric approach to texture profile analysis of kombucha fermented milk products. *Journal of Food Science and Technology is a bimonthly*, *52*(9), 5968–5974. doi:10.1007/s13197-014-1648-4

Managa, M. G., Akinola, S. A., Remize, F., Garcia, C., & Sivakumar, D. (2021). Physicochemical Parameters and Bioaccessibility of Lactic Acid Bacteria Fermented Chayote Leaf (*Sechium edule*) and Pineapple (*Ananas comosus*) Smoothies. *Frontiers in Nutrition*, *8*, 649189. doi:10.3389/fnut.2021.649189

Mandha, J., Shumoy, H., Matemu, A. O., & Raes, K. (2021). Current trends and opportunities of plant-based non-alcoholic probiotic beverages: A European and African perspective. In *Probiotic beverages* (pp. 127–158). Academic Press.

Mashitoa, F. M., Akinola, S. A., Manhevi, V. E., Garcia, C., Remize, F., Slabbert, R. M., & Sivakumar, D. (2021). Influence of fermentation of pasteurised papaya puree with different lactic acid bacterial strains on quality and bioaccessibility of phenolic compounds during in vitro digestion. *Foods*, *10*(5). doi:10.3390/foods10050962

Mbaiguinam, M., Mahmout, Y., Tarkodjiel, M., Delobel, B., & Bessiere, J.-M. (2005). Constituents of Kawal, fermented *Cassia obtusifolia* leaves, a traditional food from Chad. *African Journal of Biotechnology*, *4*(10).

Minnaar, P. P., Jolly, N. P., Paulsen, V., Du Plessis, H. W., & Van Der Rijst, M. (2017). *Schizosaccharomyces pombe* and *Saccharomyces cerevisiae* yeasts in sequential fermentations: Effect on phenolic acids of fermented Kei-apple (*Dovyalis caffra* L.) juice. *International Journal of Food Microbiology*, *257*, 232–237. doi:10.1016/j.ijfoodmicro.2017.07.004

Misihairabgwi, J., & Cheikhyoussef, A. (2017). Traditional fermented foods and beverages of Namibia. *Journal of Ethnic Foods*, *4*(3), 145–153. doi:10.1016/j.jef.2017.08.001

Montville, T. J., Matthews, K. R., & Kniel, K. E. (2012). *Food microbiology: An introduction*. ASM Press.

Mpofu, A., Linnemann, A. R., Nout, M. J., Zwietering, M. H., Smid, E. J., & den Besten, H. M. (2016). Inactivation of bacterial pathogens in yoba mutandabota, a dairy product fermented with the probiotic *Lactobacillus rhamnosus* yoba. *International Journal of Food Microbiology*, *217*, 42–48. doi:10.1016/j.ijfoodmicro.2015.09.016

Nout, M. J. R. (2014). Food technologies: Fermentation. In *Encyclopedia of food safety* (pp. 168–177). Academic Press.

Nwaiwu, O., & Itumoh, M. (2017). Chemical contaminants associated with palm wine from Nigeria are potential food safety hazards. *Beverages*, *3*(4). doi:10.3390/beverages3010016

Nyanga, L. K., Nout, M. J., Gadaga, T. H., Theelen, B., Boekhout, T., & Zwietering, M. H. (2007). Yeasts and lactic acid bacteria microbiota from masau (*Ziziphus mauritiana*) fruits and their fermented fruit pulp in Zimbabwe. *International Journal of Food Microbiology*, *120*(1–2), 159–166. doi:10.1016/j.ijfoodmicro.2007.06.021

NzaouAimé, S. A. E., & Kayath, C. K. C. (2016). *Production and Characterization of a Fibrinolytic Enzyme from Bacillus amyloliquefaciens Isolated from Ntoba Mbodi*. M.Sc. Dissertation, Marien Ngouabi University.

Odongo, G. A., Schlotz, N., Herz, C., Hanschen, F. S., Baldermann, S., Neugart, S., …, & Lamy, E. (2017). The role of plant processing for the cancer preventive potential of Ethiopian kale (*Brassica carinata*). *Food & Nutrition Research*, *61*(1), 1271527. doi:10.1080/16546628.2017.1271527

Oguntoyinbo, F. A., Fusco, V., Cho, G. S., Kabisch, J., Neve, H., Bockelmann, W., …, & Franz, C. M. (2016). Produce from Africa's Gardens: Potential for leafy vegetable and fruit fermentations. *Frontiers in Microbiology*, *7*, 981. doi:10.3389/fmicb.2016.00981

Olasupo, N., Odunfa, S., & Obayori, O. (2010). Ethnic African fermented foods. *Fermented foods and beverages of the world* (pp. 323–352). CRC Press.

Osman, N. M., Mohamed Ahmed, I. A., & Babiker, E. E. (2009). Fermentation and cooking of sicklepod (*Cassia obtusifolia*) leaves: Changes in chemical and amino acid composition, antinutrients and protein fractions and digestibility. *International Journal of Food Science & Technology*, *45*(1), 124–132. doi:10.1111/j.1365-2621.2009.02111.x

Ouoba, L., Kando, C., Parkouda, C., Sawadogo-Lingani, H., Diawara, B., & Sutherland, J. (2012). The microbiology of Bandji, palm wine of *Borassus akeassii* from Burkina Faso: Identification and genotypic diversity of yeasts, lactic acid and acetic acid bacteria. *Journal of Applied Microbiology, 113*(6), 1428–1441.

Ouoba, L. I. I., Vouidibio Mbozo, A. B., Anyogu, A., Obioha, P. I., Lingani-Sawadogo, H., Sutherland, J. P., ..., & Ghoddusi, H. B. (2019). Environmental heterogeneity of *Staphylococcus* species from alkaline fermented foods and associated toxins and antimicrobial resistance genetic elements. *International Journal of Food Microbiology, 311*, 108356. doi:10.1016/j.ijfoodmicro.2019.108356

Ouoba, L. I. I., Vouidibio Mbozo, A. B., Thorsen, L., Anyogu, A., Nielsen, D. S., Kobawila, S. C., & Sutherland, J. P. (2015). *Lysinibacillus louembei* sp. nov., a spore-forming bacterium isolated from Ntoba Mbodi, alkaline fermented leaves of cassava from the Republic of the Congo. *International Journal of Systematic and Evolutionary Microbiology, 65*(11), 4256–4262. doi:10.1099/ijsem.0.000570

Padonou, S. W., Houngbedji, M., Hounhouigan, M. H., Chadare, F. J., & Hounhouigan, D. J. (2023). B-vitamins and heat processed fermented starchy and vegetable foods in sub-Saharan Africa: A review. *Journal of Food Science, 88*(8), 3155–3188. doi:10.1111/1750-3841.16697

Parkouda, C., Ba/Hama, F., Ouattara/Songre, L., Tano-Debrah, K., & Diawara, B. (2015). Biochemical changes associated with the fermentation of baobab seeds in Maari: An alkaline fermented seeds condiment from western Africa. *Journal of Ethnic Foods, 2*(2), 58–63. doi:10.1016/j.jef.2015.04.002

Rahmani, R., Beaufort, S., Villarreal-Soto, S. A., Taillandier, P., Bouajila, J., & Debouba, M. (2019). Kombucha fermentation of African mustard (*Brassica tournefortii*) leaves: Chemical composition and bioactivity. *Food Bioscience, 30*. doi:10.1016/j.fbio.2019.100414

Rodríguez, L. G. R., Gasga, V. M. Z., Pescuma, M., Van Nieuwenhove, C., Mozzi, F., & Burgos, J. A. S. (2021). Fruits and fruit by-products as sources of bioactive compounds. Benefits and trends of lactic acid fermentation in the development of novel fruit-based functional beverages. *Food Research International, 140*, 109854.

Sanni, A. (1989). Some environmental and nutritional factors affecting growth of associated microorganisms of agadagidi. *Journal of Basic Microbiology, 29*(9), 617–622.

Saud, S., Xiaojuan, T., & Fahad, S. (2024). The consequences of fermentation metabolism on the qualitative qualities and biological activity of fermented fruit and vegetable juices. *Food Chemistry: X, 21*, 101209. doi:10.1016/j.fochx.2024.101209

Schoder, D., Maichin, A., Lema, B., & Laffa, J. (2013). Microbiological quality of milk in Tanzania: From Maasai stable to African consumer table. *Journal of Food Protection, 76*(11), 1908–1915. doi:10.4315/0362-028X.JFP-13-101

Septembre-Malaterre, A., Remize, F., & Poucheret, P. (2018). Fruits and vegetables, as a source of nutritional compounds and phytochemicals: Changes in bioactive compounds during lactic fermentation. *Food Research International*, *104*, 86–99.

Shah, N., & Singhal, R. (2017). Fermented fruits and vegetables. In: *Current developments in biotechnology and bioengineering* (pp. 45–89). Elsevier.

Swain, M. R., Anandharaj, M., Ray, R. C., & Parveen Rani, R. (2014). Fermented fruits and vegetables of Asia: A potential source of probiotics. *Biotechnology Research International*, *2014*, 250424. doi:10.1155/2014/250424

Taale, E., Gambogou, B., Sawadogo, A., Cissé, H., Souho, T., Kou'santa Amouzou, S., & Tchabi, A. (2024). Involved microorganisms in the production of indigenous fermented food from West Africa: Technological characteristics and probiotic power.

Thierry, U., & Osinachi, U. E. U. (2023). Enhancement of nutritional content of traditional banana beer using bio-process. *ASRIC Journal on Agricultural Sciences*, 107.

Ukaegbu, T., Okezie, O., & Obi, C. N. (2019). Fermentation reduces cyanide content during the production of cassava flours from sweet and bitter cassava tuber varieties. *Asian Food Science Journal*, 1–10. doi:10.9734/afsj/2019/v11i130050

Vouidibio Mbozo, A. B., Kobawila, S. C., Anyogu, A., Awamaria, B., Louembe, D., Sutherland, J. P., & Ouoba, L. I. I. (2017). Investigation of the diversity and safety of the predominant *Bacillus pumilus* sensu lato and other *Bacillus* species involved in the alkaline fermentation of cassava leaves for the production of Ntoba Mbodi. *Food Control*, *82*, 154–162. doi:10.1016/j.foodcont.2017.06.018

Watanabe, K., Fujimoto, J., Tomii, Y., Sasamoto, M., Makino, H., Kudo, Y., & Okada, S. (2009). *Lactobacillus kisonensis* sp. nov., *Lactobacillus otakiensis* sp. nov., *Lactobacillus rapi* sp. nov. and *Lactobacillus sunkii* sp. nov., heterofermentative species isolated from sunki, a traditional Japanese pickle. *International Journal of Systematic and Evolutionary Microbiology*, *59*(Pt 4), 754–760. doi:10.1099/ijs.0.004689-0

Wilson, P., David, T., & Sam, B. (2012). Microbial and biochemical changes occurring during production of traditional Rwandese banana beer "Urwagwa." *Fermentation Technology*, *1*(3). doi:10.4172/2167-7972.1000104

Yuan, X., Wang, T., Sun, L., Qiao, Z., Pan, H., Zhong, Y., & Zhuang, Y. (2024). Recent advances of fermented fruits: A review on strains, fermentation strategies, and functional activities. *Food Chemistry: X*, *22*, 101482. doi:10.1016/j.fochx.2024.101482

8

Molecular Techniques for Elucidating the Microbiology of African Fermented Products

8.1 Introduction

Elucidating the microbiology of African fermented products involves various techniques aimed at identifying and characterising the microorganisms present, monitoring species/strains, understanding their metabolic activities and assessing their impact on the fermentation process (efficiency of starter cultures) and final product quality and preservation. In food microbiology, detecting the taxonomic level of microbes depends on the sensitivity of the techniques and results may range from genus (or species) to subspecies or strain levels. Food is home to diverse microbial ecosystems, including bacteria, fungi, and viruses. Some of these bacteria are important for technology and provide food with desirable sensory qualities (rheological and organoleptic). Conversely, unfavourable microorganisms also exist and can jeopardise food safety (pathogens) or food quality (spoilage microorganisms) (Mafe et al., 2024; Quigley et al., 2011).

Conventional methods have been applied to ascertain the composition of food-associated microbiota (Figure 8.1). They depend on the isolative, cultural, morphological, physiological and biochemical characteristics of microorganisms. However, conventional methods are unreliable for a thorough microbial characterisation of food ecosystems (Mayo et al., 2014). The selective isolation of microorganisms may be hindered by unknown growth factors or by conditions that are present in natural habitats but cannot be replicated in laboratory-based experiments. Moreover, foods might be stored in harsh conditions or

DOI: 10.1201/9781003658320-8

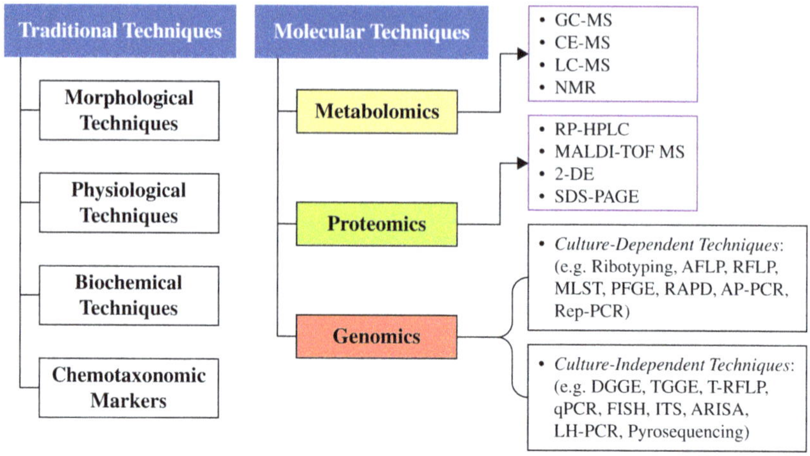

Figure 8.1 Microbial techniques used in analysing fermented foods. *Abbreviations*: GC–MS, gas chromatography–mass spectrometry; CE–MS, capillary electrophoresis–mass spectrometry; LC–MS; liquid chromatography–mass spectrometry; NMR, nuclear magnetic resonance; RP-HPLC, reversed-phase high-performance liquid chromatography; MALDI TOF MS, matrix-assisted laser desorption/ionisation time-of-flight mass spectrometry; 2-DE, two-dimensional gel electrophoresis; SDS–PAGE, sodium dodecyl sulphate-polyacrylamide gel electrophoresis; AFLP, amplified fragment length polymorphisms; RFLP, restriction fragment length polymorphism; MLST, multilocus sequence typing; PFGE, pulse field gel electrophoresis; RAPD, randomly amplified polymorphic DNA; AP-PCR, arbitrarily primed-polymerase chain reaction; Rep-PCR, Repetitive element sequence-based PCR; DGGE, denaturing gradient gel electrophoresis; TGGE, temperature gradient gel electrophoresis; T-RFLP, terminal restriction fragment length polymorphism; qPCR, quantitative-PCR; FISH, fluorescence in situ hybridisation; ITS, internal transcribed spacer; ARISA, automated ribosomal intergenic spacer analysis; LH-PCR, length heterogeneity PCR.

have a low pH, which could leave bacteria in a physiologically viable but uncultivable form. These restrictions cause such methods to underestimate the variety of microorganisms and occasionally even miss the major microbial groups (Mayo et al., 2014). This frequently presents difficulties, as it can be difficult to identify stressed cells or cultivate viable microbes. This dilemma prompted the introduction of more effective molecular techniques for studying microbial communities associated with food.

In utilising molecular identification, a precise and dependable approach integrates both culture-dependent (CD) and culture-independent (CID) techniques to identify microorganisms of African fermented foods (Banwo et al., 2019). These methodologies simplify the isolation, identification and occasionally quantification of food-associated microbial groups, offering sensitive and swift means to ascertain the composition and diversity of intricate microbial populations. Previously, traditional microbiological procedures were used to produce isolates from the food matrix for the study of nucleic acids (DNA and/or RNA), termed CD, segueing into the more recent usage of direct extraction from the food sample (CID). The scientific community widely acknowledges that CD methods often fail to adequately capture the diversity of intricate food ecosystems. This lack is primarily due to the likelihood of excluding populations present in a stressed state or with low numbers (Cocolin et al., 2011). Utilising CD methods, which depend on cultivating microorganisms, can generate an incomplete representation of the ecology of fermented food due to biases stemming from variations in the microorganisms' abilities to grow on synthetic laboratory media (Doulgeraki et al., 2017).

CID techniques are widely used in food microbiology to thoroughly profile populations of culturable and non-culturable microorganisms in fermented foods (Cocolin et al., 2013). African fermented foods are best understood through molecular methods, including PCR, 16S rRNA gene sequencing and metagenomics (Diaz et al., 2019; Ezekiel et al., 2018). An improved comprehension of microbial succession, safety and functional characteristics during fermentation has been made possible by these tools, which have also improved the identification of important functional strains and exposed hitherto uncultured and dominant microbial populations.

Over the past decade, numerous pieces of evidence have emerged, drawing attention to significant discrepancies frequently observed between the outcomes yielded by CD and CID methods. This chapter presents a thorough review of both CD and CID molecular techniques, employed to study microbial populations in various African fermented foods. The chapter further outlines the technical specifics and recent improvements of these methods.

8.2 The Need to Investigate the Microbiology of African Products

Exploring the microbiology of African food products is essential for ensuring food safety, fostering economic development and addressing public health challenges. A comprehensive understanding not only benefits local communities but also contributes to global food security and sustainability efforts (Ayalew et al., 2023). The need to explore the microbiology of African products stems from several crucial factors that underscore the importance of understanding the microbial composition and safety of these products. Ensuring the safety of food products is paramount to public health, and investigating the microbiology of African products could help identify potential pathogens or spoilage microorganisms that can pose risks to consumers if not appropriately managed. In addition, understanding the microbiology behind these processes is essential for optimising product quality, shelf life and safety, as many African products are traditionally prepared using various fermentation and preservation techniques.

African products are increasingly gaining recognition in global markets, presenting economic opportunities for producers and exporters. Investigating their microbiology can enhance product quality, consistency and compliance with international regulations, facilitating trade and economic growth. Thus, understanding the microbiology of African products enables the development of interventions and strategies to mitigate health risks and improve overall public health outcomes. This development is essential because Africa still faces various public health challenges, including infectious diseases and foodborne illnesses.

The microbial ecology of food fermentation could help create the best starting cultures and uncover indicators for evaluating the quality of fermented foods (Ghosh et al., 2023). Microbial ecology is the study of microorganisms in their natural environments; it entails investigating the various bacteria, yeasts and moulds that contribute to the fermentation processes related to food and food products (Ghosh et al., 2023). Using molecular techniques enables the identification and quantification of biomarkers, which are quantifiable indicators of specific biological states or conditions (Gbashi, Moyo, et al., 2023). In the context of fermented foods, these biomarkers may include metabolic byproducts, microbial genes, gene expression profiles, enzymatic activities or metabolite concentrations (e.g., bioactive peptides, short-chain fatty acids, polyphenol derivatives or vitamins) associated with the fermentation progress, microbial activity, pathogen detection or health-related properties of the food (Ghosh et al., 2023; Oyedeji et al., 2021). By detecting such markers, molecular methods provide insight into the microbial dynamics, functional potential and nutritional or therapeutic quality of fermented products. For example, the presence and expression levels of particular lactic acid bacterial genes or the accumulation of short-chain fatty acids, peptides and vitamins can serve as proxies for assessing the quality, safety and functional efficacy of African fermented foods (Ghosh et al., 2023; Siddiqui et al., 2023). Furthermore, the appropriate fermentation of a food product or the presence of potentially hazardous microbes can be ascertained by using specific microbial metabolites, enzymes and proteins as biomarkers (Afzaal et al., 2022).

Overall, using molecular techniques to investigate the microbiology of African products is fundamental to the selection of starter cultures for product development, preserving traditions and heritage and safeguarding public health by identifying and controlling pathogens that can cause foodborne illnesses. This approach would unlock opportunities for market expansion and international trade by ensuring compliance with stringent quality requirements.

8.3 Culture-Dependent and Culture-Independent Approaches for African Fermented Foods

8.3.1 Culture-Dependent Approaches

CD approaches are procedures involving the growth, cultivation and isolation of microbial species or strains under specific laboratory conditions based on their morphological, biochemical or physiological characteristics. These techniques are used for studying microbial diversity, characterising microbial communities and assessing the presence of pathogens in food samples (Bloomfield et al., 2023; Ercolini et al., 2001; Jany & Barbier, 2008; Wang et al., 2018). However, not all microbial strains can be recovered using randomised colony-picking techniques on culture media, and the isolates frequently fail to represent the true taxonomic diversity of microorganisms actively expressing genes accurately within the food matrix (Ndoye et al., 2011).

CD approaches are time-intensive due to the need for prolonged incubation periods, and certain species may either fail to thrive in vitro or become outcompeted by more numerically prevalent microbial species. DNA typing techniques are utilised in CD methodologies to characterise and identify samples, such as African traditional fermented foods, at a molecular level. These methods consist of pulsed-field gel electrophoresis (PFGE) for strain delineation, PCR amplification of the 16S/23S rDNA internal transcribed spacer

(ITS-PCR) for species diversity assessment, restriction fragment length polymorphism (RFLP) of the ITS-PCR for species and strain differentiation and 16S rDNA sequencing for phylogenetic relationships (Ouoba et al., 2004).

PCR targeting the ITS and repetitive element PCR have been used in studies to differentiate between Bacilli species present in *bikalga*, a vegetable condiment made from the Roselle Hibiscus plant (Oguntoyinbo et al., 2010; Ouoba et al., 2008). In another study, CD morphological analysis and 16S rRNA sequencing revealed that *Bacillus* species predominated in an *iru* sample and included *Bacillus cereus*, *Bacillus licheniformis*, *Bacillus subtilis*, *Brevibacillus formosus* and *Bacillus amyloliquefaciens*. Further characterisation using amplified ribosomal DNA restriction analysis (ARDRA), ITS-PCR, ITS-PCR-RFLP and randomly amplified polymorphic DNA (RAPD) PCR techniques demonstrated a notable diversity among the *Bacillus* isolates, with *B. amyloliquefaciens* and *B. subtilis* emerging as the most prevalent species (Adewumi et al., 2012). The authors stated that subtyping autochthonous *Bacillus* species with distinct clonal relationships and dominant groups which have been genetically characterised produced results that were consistent with earlier research on the fermentation of related vegetable proteins in West Africa (Oguntoyinbo et al., 2010; Ouoba et al., 2004; Parkouda et al., 2010).

According to Oguntoyinbo et al. (2010), the ARDRA genotyping method utilising enzymes HhaI, HinfI and Sau3AI proved ineffective in distinguishing between various *Bacillus* species, although it was effective at delineating the *B. cereus* group from others. However, combining RAPD-PCR with ARDRA genotyping methods addressed the challenges of speciation. The authors noted that a polyphasic taxonomic approach effectively separated the predominant *Bacillus* species in *okpehe* samples. In another study, *Bacillus* isolates from *soumbala* were successfully typed at the species level using ITS-PCR and ITS-PCR-RFLP, with little variation noted at the strain level (Ouoba et al., 2004). When using SfiI and SmaI as restriction enzymes to *type B. cereus, B. licheniformis* and *B. subtilis* at the strain level, PFGE proved to be the best discriminating technique for identifying *Bacillus* species from soumbala (Ouoba et al., 2004). In addition, the literature reports that both ITS-PCR and ITS-PCR-RFLP, utilising a single restriction enzyme, are easy strategies suitable for typing *Bacillus* spp. from vegetable condiments (Ouoba et al., 2004).

Research has also demonstrated the efficacy of utilising both CD and CID molecular techniques to identify dominant species, facilitating accurate identification and comprehensive study of the microbiological composition of fermented food products (Adewumi et al., 2012; Ezeokoli et al., 2016).

8.3.2 Culture-Independent Approaches

CID approaches study microbial communities by analysing the genetic material (DNA or RNA) of microorganisms directly from food samples without requiring the use of conventional laboratory culturing procedures. CID techniques have been applied to profile the dynamics, complexity, changes and evolutionary population of the microbiota of African fermented foods (Table 8.1). These techniques include quantitative PCR (qPCR), fluorescence in situ hybridisation (FISH), single-stranded conformation polymorphisms (SSCP), terminal restriction fragment length polymorphism (T-RFLP), PCR-denaturing gradient gel electrophoresis (DGGE), temporal temperature gel electrophoresis (TTGE) and DNA microarray (Adewumi et al., 2012; Oguntoyinbo et al., 2011; Solieri et al., 2012).

Since their introduction in environmental microbiology, PCR-DGGE and PCR-TGGE have become standard techniques in many laboratories globally for analysing population dynamics and composition in microbial communities associated with food. These two methods comprise amplifying the genes that encode 16S rRNA from a matrix and then separating them into DNA fragments (Cocolin & Ercolini, 2007). In particular, PCR-DGGE analysis is one of the most frequently used CID methods for identifying microorganisms in fermented foods, and this technique helps identify intra- and inter-species differences between genera (Cocolin et al., 2011; Greppi et al., 2013; Tamang, 2010; Tamang et al., 2016). Notably, the community and dynamics of microorganisms in fermented *iru* (Nigerian vegetable protein seeds) were successfully studied using PCR-DGGE (Adewumi et al., 2012). Furthermore, PCR-TTGE was employed to assess the microbiological composition of dégué and *poto poto* in a study examining how the DNA extraction method influences the community composition (Abriouel et al., 2006). Both techniques utilise a community DNA/RNA approach and are attractive because they facilitate the identification of specific species and provide a

Table 8.1 Summary of Previously Used Culture-Independent Approaches in Elucidating the Microbiology of African Fermented Foods

METHODS	RAW MATERIALS	FERMENTED FOOD	COUNTRY	TAXONOMIC RESOLUTION	REFERENCE
			PCR-BASED METHODS		
PCR-DGGE	Cereals (maize and cowpea), legumes (sorghum and millet)	Cereal/legume weaning foods	Nigeria	Community fingerprinting (genus/ species level)	Wakil et al. (2008)
	Parkia biglobosa seeds	*Iru*	Nigeria	Bacterial community profiling	Adewumi et al. (2012)
	Maize and sorghum	*Ogi*	Nigeria	Microbial communities profiling	Oguntoyinbo et al. (2011)
	Millet	*Kunu-zaki*	Nigeria	Microbial communities profiling	Oguntoyinbo et al. (2011)
	Soybean	*Soy-daddawa*	Nigeria	Microbial community structure	O. T. Ezeokoli et al. (2016)
PCR-TTGE	Maize	*Poto poto*	Congo	Bacterial community composition	Abriouel et al. (2006)
	Pearl millet	*Dengue*	Burkina Faso	Bacterial community composition	Abriouel et al. (2006)
			SEQUENCING PLATFORMS		
Shotgun sequencing	Milk	*Nunu*	Ghana	Species-level taxonomic resolution	Walsh et al. (2017)
	Milk	*Wagashi*	Benin	Taxonomic analysis and genetic relatedness	Jain et al. (2018)
16S rRNA HTS	Soybean	*Soy-daddawa*	Nigeria	Taxonomic diversity of bacterial communities	Ezeokoli et al. (2018)
	Maize	*Mahewu*	South Africa	Profiling of bacterial community	Daji et al. (2022)
	Milk	*Wara*	Nigeria	Delineation of bacterial community structure and succession pattern	Ezekiel et al. (2018)
	Sorghum	*Ogi*	Nigeria	Bacterial diversity	Diaz et al. (2019)

	Cassava	*Garri*	Ghana	Bacterial diversity	
	Maize	*Mawe*	Benin	Bacterial diversity	
	Sorghum	*Mahoto*	South Africa	Bacterial diversity	
	Millet	*Boule d'akassa*	Burkina Faso	Bacterial diversity	
	Teff	*Teff* dough	Ethiopia	Bacterial diversity	
	Honey	*Tej*	Ethiopia	Bacterial diversity	
	Sorghum	*Pito*	Ghana	Bacterial diversity	
	Palm wine	Palm tree sap	Nigeria	Bacterial diversity	
	Kwerionik	Milk	Uganda	Bacterial diversity	Diaz et al. (2019)
	Milk	*Nunu*	Ghana	Genus-level taxonomic resolution	Walsh et al. (2017)
	Milk	*Amasi*	South Africa	Bacteria composition	Maleke et al. (2022)
	Palm tree sap	Palm wine	Cote d'Ivoire	Microbial community structure	Djeni et al. (2020)
	Malted sorghum + Maize	*Umqombothi*	South Africa	Microbial community profiling	Hlangwani et al. (2023)
Pyrosequencing	Millet slurries	Ben-saalga	Burkina Faso	Microbiome diversity	Humblot and Guyot (2009)

Abbreviations: 16S rRNA HTS, 16S rRNA gene-high-throughput sequencing; PCR-DGGE, polymerase chain reaction-denaturing gradient gel electrophoresis; PCR-TTGE, PCR-temporal temperature gel electrophoresis.

comprehensive profile of how the community structure changes over time. The mechanism entails extracting nucleic acids (DNA or RNA), amplifying genes that encode 16S rRNA and analysing the resulting PCR products using a genetic fingerprinting method (Oguntoyinbo et al., 2011; Wakil et al., 2008).

Despite the many benefits of CID methods, they are not without biases, some of which result from the methods used for DNA extraction. The structural and chemical complexity of the matrices involved can make extracting DNA from food difficult (Keisam et al., 2016). Indeed, before DNA extraction, some food matrices can require pre-processing procedures (Escobar-Zepeda et al., 2016; Wei et al., 2018). Using a standardised method to assess the microbial population in various fermented samples would offer significant advantages. Although commercial extraction kits can address some bias issues, their effectiveness is often limited to commonplace food matrices and, thus, may not be suitable for complex foods (Keisam et al., 2016). To mitigate this shortcoming, Diaz et al. (2019) developed a unified amplicon sequencing method capable of analysing a broad spectrum of fermented African products spanning diverse food matrices. The authors used 16S rRNA gene amplicon sequencing to investigate the bacterial diversity of traditional fermented African products. They noted the presence of the genus Zymomonas in fermented food samples for the first time. Their findings also revealed that the *Lactobacillus* genus exhibited lower abundance in fermentations conducted under laboratory conditions compared to those conducted in artisanal or commercial settings (Diaz et al., 2019).

Many studies have employed high-throughput sequencing (HTS) techniques to investigate the microbiome of fermented foods from Africa. This has led to new discoveries regarding microbial dynamics, food safety and potential health benefits (Ezekiel et al., 2018; Obafemi et al., 2022). African researchers interested in studying the ecology of fermented foods and the microbial species/strains present therein are gradually gaining access to molecular techniques, such as whole-genome sequencing, shotgun metagenomics sequencing and 16S and 23S rRNA gene sequencing. This progression reflects the difficulties in obtaining the necessary resources and the cost implications of using HTS techniques to investigate the microbiomes and genetic diversity of strains in African fermented foods (Diaz et al.,

2019; Obafemi et al., 2022). HTS allows for a detailed description of food-related microbial isolates and simplifies the analysis of mixed microbial communities (Walsh et al., 2017).

Several HTS platforms, including those developed by *Illumina, Oxford Nanopore, Ion Torrent, Pacific Biosystems* and their respective protocols, have the capability to unveil microbial taxonomy, diversity, functional potential and gene expression patterns. HTS can be a first step in comprehending the ecology of foods and providing information for the targeted isolation of microorganisms, even though it cannot determine the viability and count of microbial cells in a particular food matrix (Ezeokoli et al., 2018). The techniques employed to examine fermented foods in Africa have often hindered the characterisation of so-called uncultivable species or uncommon microbes. This can be circumvented by developing modified DNA extraction protocols for the subsequent genomic and metagenomic sequencing of specific African fermented foods such as *fura* (Ghana), *masau* (Zimbabwe), obusera (Uganda), *lafun* (Nigeria), *ogi* (Nigeria) and *dengue* (Burkina Faso) (Obafemi et al., 2022).

Metagenomic sequencing, which could be either sequence- or function-based, facilitates the CID exploration of complex microbial communities by assembling metagenome-derived genomes. Generally, a standard metagenomic sequencing process entails nucleic acid extraction and purification, library preparation, sequencing and subsequent data analysis through a bioinformatics pipeline (Figure 8.2) (Ambardar et al., 2016). Two commonly employed applications in metagenomics studies and microbial profiling are 16S rRNA analysis and shotgun metagenomics. The 16S rRNA analysis relies on distinguishing subunits of prokaryotic and eukaryotic ribosomes, selectively amplifying the conserved hypervariable regions (V1–V9) within bacterial cells, thereby achieving high-fidelity sequences with minimal errors. In contrast, shotgun metagenomics encompasses sequencing the entire genomic DNA, offering broader genomic coverage and higher data output than 16S rRNA sequencing (Jeong et al., 2021). The application of shotgun metagenomic sequencing is on the rise to decipher the composition of microbial populations involved in food fermentations (Verce et al., 2019). To date, from an African standpoint, it has been utilised to examine the mechanisms of spontaneous fermentation of nunu dairy beverages (Walsh et al., 2017).

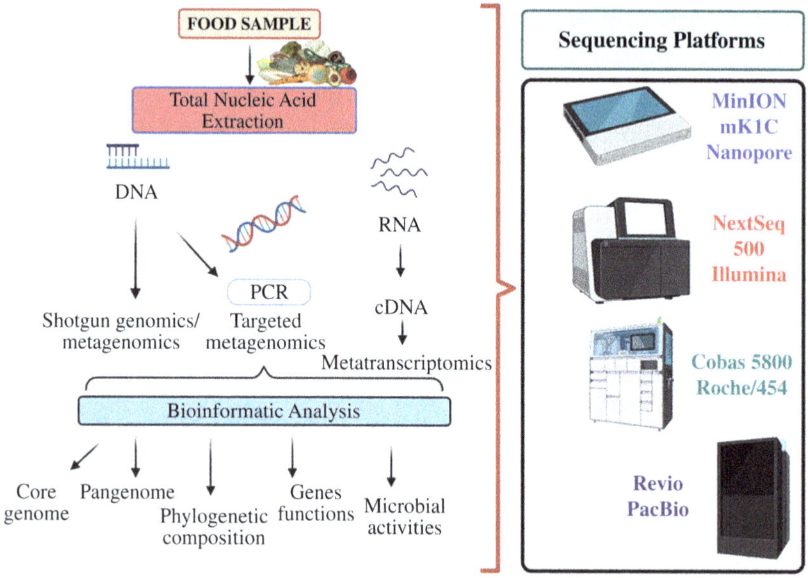

Figure 8.2 Approach to the microbial analysis of fermented food samples using metagenomic sequencing techniques.

At present, three generations of DNA sequencing techniques can be employed to elucidate the microbiology of African fermented foods (Gbashi, Tebele, et al., 2023). First-generation methods employ the chain termination approach, achieving read lengths of 500–1000 bp. The second-generation techniques encompass sequencing by synthesis or ligation, with read lengths ranging from 50 to 500 bp. Next-generation sequencing (NGS) discovery represents a significant leap forward in examining biological systems, enabling the development of highly efficient and feasible research tools for exploring intricate genomes that prove challenging or unfeasible to analyse with traditional sequencing approaches. The emergence of NGS technologies has created a distinct platform for investigating the microbial ecology of any food environment with unparalleled depth and coverage. Such examination enables a deeper understanding of the microbial ecology within food matrices previously studied (Ezeokoli et al., 2018). Third-generation sequencing technologies conduct single-molecule sequencing with read lengths measured in tens of kilobases (kb) (Akacin et al., 2022).

8.4 Future Trends for Molecular Techniques in the Microbiology of African Fermented Foods

A growing number of consumers across Africa are becoming better informed and more conscious of the beneficial effects of fermented foods on health and overall wellness, which highlights a significant opportunity for advancing research on fermented foods. Exploiting this opportunity would necessitate adopting a unified systems approach, allowing for a deeper understanding of how specific microorganisms present in various African regions contribute to the fermentation processes of foods. In turn, such understanding enhances the functionality of these foods to accommodate the diverse dietary needs of populations across the continent (Banwo et al., 2019). Chapter 8 expounded on the significance of microorganisms in African fermented foods and the role of various molecular techniques in their identification. The chapter advocated for adopting an approach that integrates phenotypic and multiple genotypic methods, which could facilitate the exploration of the functionality of these microorganisms as starter cultures in traditional fermented foods across Africa (Banwo et al., 2019; Diaz et al., 2019; Obafemi et al., 2022).

The integration of both CD and CID techniques is essential for characterising microorganisms, understanding the dynamics and exploring the diversities inherent in the fermentation processes of indigenous fermented foods (Adewumi et al., 2012; Oguntoyinbo et al., 2011). This is crucial for gaining insights into their behaviours and activities as well as the attributes they impart to fermented food products, encompassing aspects such as appearance, taste, nutritional value and safety.

A collaborative effort among African researchers is also essential for preserving and utilising microorganisms isolated from indigenous fermented foods. These microorganisms should be adequately characterised and identified, with strains exhibiting distinct properties to undergo complete genome sequencing for ongoing research purposes. Establishing a dedicated microbial culture collection centre would facilitate their classification based on their unique features, enabling easy access for research endeavours. This initiative would streamline the acquisition, validation, propagation, preservation and dissemination of microbial information for scientific research within the African region (Adewumi et al., 2012; Adewumi et al., 2014; Banwo et al., 2019; Diaz et al., 2019; Djeni et al., 2020; Oguntoyinbo et al., 2011).

The application of polyphasic taxonomic techniques has proved invaluable in identifying strains and gaining a comprehensive understanding of the microbial diversity present in various indigenous fermented foods in Africa (Adewumi et al., 2012; Adewumi et al., 2014; Banwo et al., 2019; Diaz et al., 2019; Djeni et al., 2020; Oguntoyinbo et al., 2011). In future, a strong CID approach for investigating fermented food samples could be achieved with high throughput based on 'sequencing by synthesis' to profile microbial communities (Akacin et al., 2022). This method is not labour-intensive and obviates the necessity of cloning and CD techniques, thus circumventing the generation of aberrant recombinants and artefacts associated with cloning (Ndoye et al., 2011). Moreover, this would improve our understanding of the flavour, taste and texture of foods that have undergone fermentation (Siddiqui et al., 2023).

African researchers could also employ new approaches in metagenomics, which involves the direct sequencing of DNA extracted from a sample without culturing. It allows for a comprehensive analysis of all microbial DNA present in a sample, providing information about the diversity and abundance of microorganisms (Billington et al., 2022). In future, metagenomics will likely become more accessible and cost-effective, allowing researchers to gain deeper insights into the complex microbial ecosystems that play a role in African fermented foods. In addition, integrating data from genomics, transcriptomics, proteomics and metabolomics would provide a holistic view of microbial communities and their interactions, thereby helping unravel the complex networks of metabolic pathways and regulatory mechanisms underlying fermentation processes. Even functional genomics techniques, such as metatranscriptomics and metaproteomics, could reveal the gene expression patterns and protein profiles of microbial communities in fermented foods. Future research could focus on linking these molecular profiles to specific fermentation processes, flavour development and nutritional properties of African fermented foods.

By employing these newer techniques either individually or in combination, researchers could gain a comprehensive understanding of the microbiology of African fermented food products, optimising the fermentation process and improving product quality and safety. However, while some of the techniques mentioned here may be more readily available in certain African regions or institutions with well-established research programmes, other regions might be limited by cost or infrastructure constraints. For example, the polymerase chain reaction can be relatively affordable compared to more advanced molecular biology equipment, but the cost of PCR consumables and reagents can be a significant challenge, especially for laboratories with limited budgets. DNA sequencing and techniques for microbial community profiling can be expensive, particularly for HTS technologies, such as NGS of 16S rRNA or ITS regions. Collaborations with international research partners or utilising sequencing services provided by commercial companies might be options for accessing sequencing capabilities. Metagenomic analysis, typically involving DNA extraction, library preparation, sequencing and bioinformatics analysis, can also incur significant costs. Metagenomic analysis requires access to sequencing facilities with the capacity for HTS (Akacin et al., 2022), and such facilities may not be widely available across the continent. Thus, initiatives focused on technology transfer, capacity building, research collaborations and reducing the cost of reagents and equipment could increase accessibility to these techniques across the continent (Diaz et al., 2019).

8.5 Conclusion

Exploring techniques for elucidating the microbiology of African fermented products represents a dynamic and evolving field with significant implications for food science, food chemistry and microbiology. Diverse molecular tools enable researchers to more extensively examine the microbial diversity, metabolic activities and functional attributes of microorganisms involved in African fermented foods. However, it is crucial to recognise the importance of considering cultural, socioeconomic and ecological dimensions in studying African fermented foods. Thus, interdisciplinary collaborations between scientists, food technologists, anthropologists and local communities are essential for conducting research that respects and preserves cultural heritage while promoting food safety, quality and sustainability. The continued advancement of molecular techniques in understanding the microbiology of African fermented products holds promise for fostering innovation, promoting food security and supporting the sustainable development of diverse food cultures.

References

Abriouel, H., Ben Omar, N., Lopez, R. L., Martinez-Canamero, M., Keleke, S., & Galvez, A. (2006). Culture-independent analysis of the microbial composition of the African traditional fermented foods poto poto and degue by using three different DNA extraction methods. *International Journal of Food Microbiology*, *111*(3), 228–233. doi:10.1016/j.ijfoodmicro.2006.06.006

Adewumi, G. A., Oguntoyinbo, F. A., Keisam, S., Romi, W., & Jeyaram, K. (2012). Combination of culture-independent and culture-dependent molecular methods for the determination of bacterial community of iru, a fermented *Parkia biglobosa* seeds. *Frontiers in Microbiology*, *3*, 436. doi:10.3389/fmicb.2012.00436

Adewumi, G. A., Oguntoyinbo, F. A., Romi, W., Singh, T. A., & Jeyaram, K. (2014). Genome subtyping of autochthonous *Bacillus* species isolated from Iru, a fermented *Parkia biglobosa* seed. *Food Biotechnology*, *28*(3), 250–268. doi:10.1080/08905436.2014.931866

Afzaal, M., Saeed, F., Hussain, M., Shahid, F., Siddeeg, A., & Al-Farga, A. (2022). Proteomics as a promising biomarker in food authentication, quality and safety: A review. *Food Science & Nutrition*, *10*(7), 2333–2346. doi:10.1002/fsn3.2842

Akacin, I., Ersoy, S., Doluca, O., & Gungormusler, M. (2022). Comparing the significance of the utilization of next generation and third generation sequencing technologies in microbial metagenomics. *Microbiological Research*, *264*, 127154. doi:10.1016/j.micres.2022.127154

Ambardar, S., Gupta, R., Trakroo, D., Lal, R., & Vakhlu, J. (2016). High throughput sequencing: An overview of sequencing chemistry. *Indian Journal of Microbiology*, *56*(4), 394–404. doi:10.1007/s12088-016-0606-4

Ayalew, A., Kareem, F., & Randolph, D. (2023). A paradigm shift in food safety for Africa. In Ulimwengu, J. M., Kwofie, E. M. & Collins, J. (Eds), *African Food Systems Transformation and the Post-Malabo Agenda, ReSAKSS Annual Trends and Outlook Report*, (pp. 106–120), AKADEMIYA2063 and International Food Policy Research Institute (IFPRI).

Banwo, K., Adesulu-Dahunsi, A. T., & Dahunsi, S. O. (2019). Molecular methods in fermented foods. In: *Fermented food products* (pp. 365–382). CRC Press.

Billington, C., Kingsbury, J. M., & Rivas, L. (2022). Metagenomics approaches for improving food safety: A review. *Journal of Food Protection*, *85*(3), 448–464. doi:10.4315/JFP-21-301

Bloomfield, S. J., Zomer, A. L., O'Grady, J., Kay, G. L., Wain, J., Janecko, N., …, & Mather, A. E. (2023). Determination and quantification of microbial communities and antimicrobial resistance on food through host DNA-depleted metagenomics. *Food Microbiology, 110*, 104162. doi:10.1016/j.fm.2022.104162

Cocolin, L., Alessandria, V., Dolci, P., Gorra, R., & Rantsiou, K. (2013). Culture independent methods to assess the diversity and dynamics of microbiota during food fermentation. *International Journal of Food Microbiology, 167*(1), 29–43. doi:10.1016/j.ijfoodmicro.2013.05.008

Cocolin, L., Dolci, P., & Rantsiou, K. (2011). Biodiversity and dynamics of meat fermentations: The contribution of molecular methods for a better comprehension of a complex ecosystem. *Meat Sci, 89*(3), 296–302. doi:10.1016/j.meatsci.2011.04.011

Cocolin, L., & Ercolini, D. (2007). *Molecular techniques in the microbial ecology of fermented foods*. Springer Science & Business Media.

Daji, G. A., Green, E., Abrahams, A., Oyedeji, A. B., Masenya, K., Kondiah, K., & Adebo, O. A. (2022). Physicochemical properties and bacterial community profiling of optimal Mahewu (A fermented food product) prepared using white and yellow maize with different inocula. *Foods, 11*(20). doi:10.3390/foods11203171

Diaz, M., Kellingray, L., Akinyemi, N., Adefiranye, O. O., Olaonipekun, A. B., Bayili, G. R., …, & Narbad, A. (2019). Comparison of the microbial composition of African fermented foods using amplicon sequencing. *Scientific Reports, 9*(1), 13863. doi:10.1038/s41598-019-50190-4

Djeni, T. N., Kouame, K. H., Ake, F. D. M., Amoikon, L. S. T., Dje, M. K., & Jeyaram, K. (2020). Microbial diversity and metabolite profiles of palm wine produced from three different palm tree species in cote d'Ivoire. *Scientific Reports, 10*(1), 1715. doi:10.1038/s41598-020-58587-2

Doulgeraki, A., Panagou, E., & Nychas, G. J. (2017). Rapid methods for microbial analysis of meat and meat products. *Emerging technologies in meat processing: Production, processing and technology* (pp. 321–344). Wiley.

Ercolini, D., Moschetti, G., Blaiotta, G., & Coppola, S. (2001). The potential of a polyphasic PCR-dGGE approach in evaluating microbial diversity of natural whey cultures for water-buffalo Mozzarella cheese production: Bias of culture-dependent and culture-independent analyses. *Systematic and Applied Microbiology, 24*(4), 610–617. doi:10.1078/0723-2020-00076

Escobar-Zepeda, A., Sanchez-Flores, A., & Quirasco Baruch, M. (2016). Metagenomic analysis of a Mexican ripened cheese reveals a unique complex microbiota. *Food Microbiology, 57*, 116–127. doi:10.1016/j.fm.2016.02.004

Ezekiel, C. N., Ayeni, K. I., Ezeokoli, O. T., Sulyok, M., van Wyk, D. A. B., Oyedele, O. A., …, & Krska, R. (2018). High-throughput sequence analyses of bacterial communities and multi-mycotoxin profiling during processing of different formulations of Kunu, a traditional fermented beverage. *Frontiers in Microbiology, 9*, 3282. doi:10.3389/fmicb.2018.03282

Ezeokoli, O., Gupta, A., Popoola, T., & Bezuidenhout, C. (2016). Molecular analysis of bacterial community dynamics during the fermentation of soy-daddawa condiment. *Food Science and Biotechnology, 25*(4), 1081–1086. doi:10.1007/s10068-016-0174-8

Ezeokoli, O. T., Adeleke, R. A., & Bezuidenhout, C. C. (2018). Core bacterial community of soy-daddawa: Insights from high-throughput DNA metabarcoding. *Lwt*, *97*, 61–66. doi:10.1016/j.lwt.2018.06.039

Ezeokoli, O. T., Gupta, A. K., Mienie, C., Popoola, T. O., & Bezuidenhout, C. C. (2016). PCR-denaturing gradient gel electrophoresis analysis of microbial community in soy-daddawa, a Nigerian fermented soybean (Glycine max (L.) Merr.) condiment. *International Journal of Food Microbiology*, *220*, 58–62. doi:10.1016/j.ijfoodmicro.2016.01.003

Gbashi, S., Moyo, S. M., Olopade, B., Kewuyemi, Y., Areo, O. M., Lawal, O. M., Momoh, C. O., Igbashio, M. D., & Njobeh, P. B. (2023). African Fermented vegetable and fruit-based products. In *Indigenous fermented foods for the tropics* (pp. 227–244). Academic Press.

Gbashi, S., Tebele, S. M., & Njobeh, P. B. (2023). Metagenomics for the identification and characterization of microorganisms in fermented foods. In *Indigenous fermented foods for the tropics* (pp. 347–359). Academic Press.

Ghosh, S., Bornman, C., Meskini, M., & Joghataei, M. (2023). Microbial diversity in African foods and beverages: A systematic assessment. *Current Microbiology*, *81*(1), 19. doi:10.1007/s00284-023-03481-z

Greppi, A., Rantsiou, K., Padonou, W., Hounhouigan, J., Jespersen, L., Jakobsen, M., & Cocolin, L. (2013). Determination of yeast diversity in ogi, mawe, gowe and tchoukoutou by using culture-dependent and -independent methods. *International Journal of Food Microbiology*, *165*(2), 84–88. doi:10.1016/j.ijfoodmicro.2013.05.005

Hlangwani, E., Abrahams, A., Masenya, K., & Adebo, O. A. (2023). Analysis of the bacterial and fungal populations in South African sorghum beer (umqombothi) using full-length 16S rRNA amplicon sequencing. *World Journal of Microbiology and Biotechnology*, *39*(12), 350. doi:10.1007/s11274-023-03764-4

Humblot, C., & Guyot, J. P. (2009). Pyrosequencing of tagged 16S rRNA gene amplicons for rapid deciphering of the microbiomes of fermented foods such as pearl millet slurries. *Applied and Environmental Microbiology*, *75*(13), 4354–4361. doi:10.1128/AEM.00451-09

Jain, C., Rodriguez, R. L., Phillippy, A. M., Konstantinidis, K. T., & Aluru, S. (2018). High throughput ANI analysis of 90K prokaryotic genomes reveals clear species boundaries. *Nature Communications*, *9*(1), 5114. doi:10.1038/s41467-018-07641-9

Jany, J. L., & Barbier, G. (2008). Culture-independent methods for identifying microbial communities in cheese. *Food Microbiology*, *25*(7), 839–848. doi:10.1016/j.fm.2008.06.003

Jeong, J., Yun, K., Mun, S., Chung, W. H., Choi, S. Y., Nam, Y. D., Lim, M. Y., Hong, C. P., Park, C., Ahn, Y. J., & Han, K. (2021). The effect of taxonomic classification by full-length 16S rRNA sequencing with a synthetic long-read technology. *Scientific Reports*, *11*(1), 1727. doi:10.1038/s41598-020-80826-9

Keisam, S., Romi, W., Ahmed, G., & Jeyaram, K. (2016). Quantifying the biases in metagenome mining for realistic assessment of microbial ecology of naturally fermented foods. *Scientific Reports*, *6*, 34155. doi:10.1038/srep34155

Mafe, A. N., Edo, G. I., Makia, R. S., Joshua, O. A., Akpoghelie, P. O., Gaaz, T. S., Jikah, A. N., Yousif, E., Isoje, E. F., Igbuku, U. A., Ahmed, D. S., Essaghah, A. E. A., & Umar, H. (2024). A review on food spoilage mechanisms, food borne diseases and commercial aspects of food preservation and processing. *Food Chemistry Advances*, *5*. doi:10.1016/j.focha.2024.100852

Maleke, M., Doorsamy, W., Abrahams, A. M., Adefisoye, M. A., Masenya, K., & Adebo, O. A. (2022). Influence of fermentation conditions (temperature and time) on the physicochemical properties and bacteria Microbiota of Amasi. *Fermentation*, *8*(2). doi:10.3390/fermentation8020057

Mayo, B., TCC Rachid, C., Alegría, Á, MO Leite, A., S Peixoto, R., & Delgado, S. (2014). Impact of next generation sequencing techniques in food microbiology. *Current Genomics*, *15*(4), 293–309.

Ndoye, B., Rasolofo, E. A., LaPointe, G., & Roy, D. (2011). A review of the molecular approaches to investigate the diversity and activity of cheese microbiota. *Dairy Science & Technology*, *91*(5), 495–524. doi:10.1007/s13594-011-0031-8

Obafemi, Y. D., Oranusi, S. U., Ajanaku, K. O., Akinduti, P. A., Leech, J., & Cotter, P. D. (2022). African fermented foods: Overview, emerging benefits, and novel approaches to microbiome profiling. *npj Science of Food*, *6*(1), 15. doi:10.1038/s41538-022-00130-w

Oguntoyinbo, F. A., Huch, M., Cho, G. S., Schillinger, U., Holzapfel, W. H., Sanni, A. I., & Franz, C. M. (2010). Diversity of bacillus species isolated from okpehe, a traditional fermented soup condiment from Nigeria. *Journal of Food Protection*, *73*(5), 870–878. doi:10.4315/0362-028x-73.5.870

Oguntoyinbo, F. A., Tourlomousis, P., Gasson, M. J., & Narbad, A. (2011). Analysis of bacterial communities of traditional fermented West African cereal foods using culture independent methods. *International Journal of Food Microbiology*, *145*(1), 205–210. doi:10.1016/j.ijfoodmicro.2010.12.025

Ouoba, L. I., Diawara, B., Amoa-Awua, W., Traore, A. S., & Moller, P. L. (2004). Genotyping of starter cultures of *Bacillus subtilis* and *Bacillus pumilus* for fermentation of African locust bean (*Parkia biglobosa*) to produce Soumbala. *International Journal of Food Microbiology*, *90*(2), 197–205. doi:10.1016/s0168-1605(03)00302-7

Ouoba, L. I., Parkouda, C., Diawara, B., Scotti, C., & Varnam, A. H. (2008). Identification of *Bacillus* spp. from Bikalga, fermented seeds of *Hibiscus sabdariffa*: Phenotypic and genotypic characterization. *Journal of Applied Microbiology*, *104*(1), 122–131. doi:10.1111/j.1365-2672.2007.03550.x

Oyedeji, A. B., Green, E., Adebiyi, J. A., Ogundele, O. M., Gbashi, S., Adefisoye, M. A., …, & Adebo, O. A. (2021). Metabolomic approaches for the determination of metabolites from pathogenic microorganisms: A review. *Food Research International*, *140*, 110042. doi:10.1016/j.foodres.2020.110042

Parkouda, C., Thorsen, L., Compaore, C. S., Nielsen, D. S., Tano-Debrah, K., Jensen, J. S., Diawara, B., & Jakobsen, M. (2010). Microorganisms associated with Maari, a baobab seed fermented product. *International Journal of Food Microbiology*, *142*(3), 292–301. doi:10.1016/j.ijfoodmicro.2010.07.004

Quigley, L., O'Sullivan, O., Beresford, T. P., Ross, R. P., Fitzgerald, G. F., & Cotter, P. D. (2011). Molecular approaches to analysing the microbial composition of raw milk and raw milk cheese. *International Journal of Food Microbiology*, *150*(2–3), 81–94. doi:10.1016/j.ijfoodmicro.2011.08.001

Siddiqui, S. A., Erol, Z., Rugji, J., Taşçı, F., Kahraman, H. A., Toppi, V., Musa, L., Di Giacinto, G., Bahmid, N. A., Mehdizadeh, M., & Castro-Muñoz, R. (2023). An overview of fermentation in the food industry – Looking back from a new perspective. *Bioresources and Bioprocessing*, *10*(1). doi:10.1186/s40643-023-00702-y

Solieri, L., Dakal, T. C., & Giudici, P. (2012). Next-generation sequencing and its potential impact on food microbial genomics. *Annals of Microbiology*, *63*(1), 21–37. doi:10.1007/s13213-012-0478-8

Tamang, J. P. (2010). Diversity of fermented foods. *Fermented Foods and Beverages of the World*, *1*, 41–83.

Tamang, J. P., Watanabe, K., & Holzapfel, W. H. (2016). Review: Diversity of microorganisms in global fermented foods and beverages. *Frontiers in Microbiology*, *7*, 377. doi:10.3389/fmicb.2016.00377

Verce, M., De Vuyst, L., & Weckx, S. (2019). Shotgun metagenomics of a water kefir fermentation ecosystem reveals a novel oenococcus species. *Frontiers in Microbiology*, *10*, 479. doi:10.3389/fmicb.2019.00479

Wakil, S., Onilude, A., Adetutu, E., & Ball, A. (2008). PCR-DGGE fingerprints of microbial successional changes during fermentation of cereal-legume weaning foods. *African Journal of Biotechnology*, *7*(24).

Walsh, A. M., Crispie, F., Daari, K., O'Sullivan, O., Martin, J. C., Arthur, C. T., Claesson, M. J., Scott, K. P., & Cotter, P. D. (2017). Strain-level metagenomic analysis of the fermented dairy beverage nunu highlights potential food safety risks. *Applied and Environmental Microbiology*, *83*(16). doi:10.1128/AEM.01144-17

Wang, X., Ren, H., & Zhan, Y. (2018). Characterization of microbial community composition and pathogens risk assessment in typical Italian-style salami by high-throughput sequencing technology. *Food Science and Biotechnology*, *27*(1), 241–249. doi:10.1007/s10068-017-0200-5

Wei, Y. J., Wu, Y., Yan, Y. Z., Zou, W., Xue, J., Ma, W. R., Wang, W., Tian, G., & Wang, L. Y. (2018). High-throughput sequencing of microbial community diversity in soil, grapes, leaves, grape juice and wine of grapevine from China. *PLoS One*, *13*(3), e0193097. doi:10.1371/journal.pone.0193097

9

Emerging Omics for Use in African Fermented Foods

9.1 Introduction

African diets include fermented foods because they are essential for nutrition, cultural practices and food preservation. The preparation techniques of these African fermented foods (highlighted in Chapters 1–7) have been handed down through generations, during which time the quality and integrity of such products were measured through folklore or empirical knowledge. With the aid of cutting-edge scientific instruments, especially omics technologies, traditional fermentation methods that were previously primarily informed by folklore are now being re-examined. The term omics encompasses a wide range of high-throughput screening (HTS) techniques, including genomics, transcriptomics, metabolomics and proteomics. The *omics* domain offers unprecedented insights into the metabolic processes, microbial communities and functional properties of fermented foods. A metabolic and functional understanding of microbial populations and their effects on fermented products, including the compounds that impart flavour and aroma as well as their nutritional value, is possible through the application of omics science to the study of fermented foods. The application of omics technology, specifically in food science research, is known as *foodomics*, combining the words food and omics (Adebo et al., 2021; Cifuentes, 2009).

The use of HTS techniques to investigate African fermented foods is not unusual; for example, studies using gas chromatography–mass spectrometry (GC–MS) to identify volatile compounds and the aroma profiles of the fermented milk, *amasi* (Gadaga et al., 2007), palm wine, *nsamba*, retted cassava dough, *bikedi* (Dhellot

DOI: 10.1201/9781003658320-9

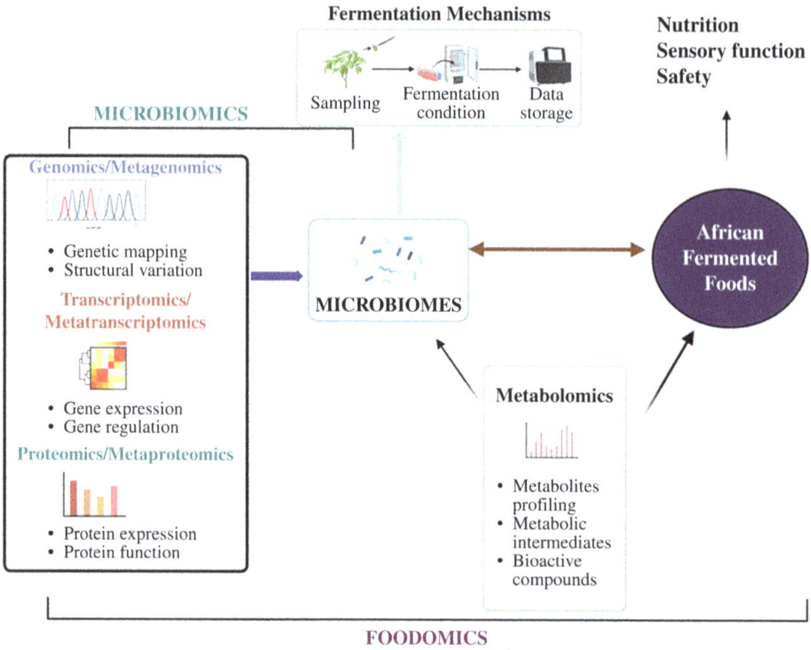

Figure 9.1 General schematic overview of the omics approach in the research of African fermented foods.

et al., 2014) and condiments, *dawadawa*, *iru* and *sonru* (Akanni, De Kock, et al., 2018; Azokpota et al., 2008; Ouoba et al., 2005). However, the scope of these studies was limited to identifying the aroma profiles of the fermented product compared to the versatility of the techniques used. In Chapter 8, advances in the use of genomics for microbial community profiling in African fermented foods were discussed.

Transcriptomics, proteomics and metabolomics are other omics methods that can be applied to identify fundamental molecular food mechanisms in connection to health and nutrition (Figure 9.1) (Li et al., 2021). For instance, metagenomics allows for identifying previously unculturable microorganisms, while metabolomics can track biochemical changes during fermentation, offering opportunities to optimise processes for better health outcomes and shelf stability (Adebo et al., 2021). Transcriptomic studies are increasingly vital in food microbiology and are relevant to understanding microbial responses and interactions in traditional fermented foods (Tamang et al., 2020). Table 9.1 provides a summary of omics techniques used in studies on African fermented products. Applying omics in African

Table 9.1 Summary of Some Omics Technologies Used on African Fermented Products

PRODUCT	COUNTRY	AIM OF STUDY	IDENTIFICATION/ ANALYTICAL PLATFORM	REFERENCES
Poto-Poto	Congo	Examine the molecular processes that go into producing *poto-poto*	MALDI-TOF/MS	Turpin et al. (2010)
Fufu	Southern Nigeria	An effort to increase the nutritional content of *fufu* flour	Proximate composition	Olapade et al. (2014)
Kenkey	Ghana	Examine the molecular processes that go into producing *kenkey*	MALDI-TOF/MS	Turpin et al. (2010)
Gari	West Africa	To improve the nutritional value of cassava products (*gari* and flour)	Proximate composition	Oboh and Akindahunsi (2003) and Oboh et al. (2002)
Dawadawa	Nigeria	Characterisation of the microbial communities responsible for fermentation	MALDI-TOF/MS	Akanni, Naudé, et al. (2018)
Kunu-Zaki/ Kunu Aya	Nigeria	Profiling of the lactic acid bacteria (LAB)	MALDI-TOF/MS	Ogunremi et al. (2022) and Ogunremi et al. (2024)
Baca or Wômi	Côte d'Ivoire	Identification of LAB strains	MALDI-TOF/MS	Soro-Yao et al. (2014)
Ting	Southern Africa	Profiling metabolites to decipher the intricate biochemical changes that occur during fermentation	GC-HRTOF-MS	Adebo et al. (2019)
Ogi	Nigeria	Metabolite profiling during fermentation	GC-HRTOF-MS	Odukoya et al. (2023)
Mahewu	South Africa	To investigate metabolite changes and biochemical alterations in maize-based fermented *mahewu* using different inocula	GC-HRTOF-MS	Akanni et al. (2025) and Daji et al. (2024)

Abbreviations: GC-HRTOF-MS, Gas Chromatography coupled with High-Resolution Time-of-Flight Mass Spectrometry; MALDI-TOF/MS, Matrix-Assisted Laser Desorption/Ionisation Time-of-Flight Mass Spectrometry.

fermented food science not only enhances scientific understanding but also supports innovation in food product development, food security and public health. This section highlights the emerging use of transcriptomics, metabolomics, proteomics and multi-omics platforms in terms of investigating African fermented foods.

9.2 Transcriptomics

Transcriptomics is a subset of the field of omics sciences, which studies the functional features of a genome and its molecular components. It comprises all aspects of RNA, such as expression levels, roles, locations and degradation, and has thus paved the way for a better understanding of how genes are expressed and linked (Wang et al., 2009). Transcriptomics is defined as the study of the whole set of RNA transcripts (the transcriptome) synthesised by an organism in a specific condition. It provides crucial information on microbial activity during food fermentation. Unlike metagenomics, transcriptomics enables the understanding of active metabolism through the measurement of gene expression levels (Wang et al., 2009). Transcriptomic studies are now gaining prominence as essential methods in food microbiology, specifically for the interpretation of microbial reactions to environmental states, inter-species interactions and substrate use. These applications are relevant to the study of traditional African fermented foods, which are commonly produced via spontaneous fermentation processes that enhance nutritional improvement, food safety, preservation ease and sensory qualities (Tamang et al., 2020).

Recent developments in (meta)transcriptomic techniques have yielded novel knowledge about the microbial communities executing these processes and improved our insights into their functional roles, with the possibility of optimisation. Metatranscriptomic approaches, such as RNA sequencing (RNA-seq) and quantitative real-time polymerase chain reaction (qRT-PCR), are employed to investigate the dynamic gene expression patterns of microbial communities in fermented foods (Sripathi et al., 2021; Yang et al., 2022). The workflow normally starts with the extraction of high-quality RNA from intricate fermentative samples, which ultimately leads to the generation of cDNA libraries using either rRNA depletion or mRNA enrichment based on the prokaryotic or eukaryotic characteristics of the targeted transcripts. High-throughput sequencing, most commonly via *Illumina* platforms, generates massive datasets that are processed through quality control, read alignment (e.g., using *HISAT2* or *STAR*) and expression quantification tools such as *FeatureCounts* or *HTSeq* (Conesa et al., 2016; Satam et al., 2023). In instances where reference genomes are not available, as is often the case with poorly described native microbes, the application of *de novo* transcriptome assembly software is required. These methods enable comparison of differential gene expression (e.g., by *DESeq2* or *edgeR*), thus identifying active genes during varying fermentation conditions and interactions among microbes (Conesa et al., 2016; Wang et al., 2009).

Jung et al. (2013) used metatranscriptomics to examine microbial gene expression profiles during *kimchi* (a traditional food product) fermentation. The authors analysed six lactic acid bacteria (LAB) strains, including *Leuconostoc mesenteroides, Latilactobacillus sakei, Weissella koreensis, Leuconostoc gelidum*, *Leuconostoc carnosum* and *Leuconostoc gasicomitatum*, using available reference genomes from Leu at five different time points throughout a 29-day fermentation. After mapping, the researchers discovered that *L. sakei* and *W. koreensis* had high gene expression at later stages of fermentation, whereas *L. mesenteroides* was the most active during the early stages. The study found that several genes associated with carbohydrate transport, hydrolysis and lactate fermentation were highly expressed, suggesting heterolactic acid fermentation. Genes encoding mannitol dehydrogenase (*mdh*) were detected in all *Leuconostoc* species, with *L. mesenteroides* notably containing three copies of the *mdh* gene (two on the chromosome and one on the plasmid), each exhibiting distinct expression patterns (Jung et al., 2013). Another study also employed transcriptomic analysis to investigate the metabolic activities of *L. mesenteroides* during *kimchi* fermentation (Chun et al., 2017). The mRNA sequencing data revealed that *L. mesenteroides* was most active in the early fermentation stages, with relative transcript abundances declining by day 18. KEGG pathway mapping indicated high expression of genes involved in carbohydrate metabolism, nucleotide metabolism, fatty acid biosynthesis, oxidative phosphorylation, riboflavin synthesis and glutamine/glutamate metabolism. Further analysis revealed that *L. mesenteroides* actively induced heterolactic fermentation by utilising a wide range of carbohydrates. This result is due to a high expression of genes involved in the production of lactate, ethanol and CO_2, as well as sugar transporter genes for glucose, fructose and mannose.

Lessard et al. (2014) investigated another traditional food product, a Camembert-type cheese, the ripening of which is often driven by fungal species such as *Geotrichum candidum* and *Penicillium camemberti* (Lessard et al., 2014). This study analysed the metatranscriptome profile for over 77 days, generating a database (CamemBank01) with over a million sequence tags assembled in 7916 contigs. The results revealed that moulds dominated the fungal community and key ripening processes, including gene expression, metabolism and stress responses, primarily occurring in the first two weeks.

In African food fermentation, where spontaneous, mixed-species fermentations frequently occur, (meta) transcriptomics can provide insight into key functional characteristics and stress behaviour of dominant microorganisms. This information is needed for the selection or design of robust starter cultures with desired technological and safety profiles. Moreover, integrating transcriptomic data with other omics data, such as metabolomics and proteomics, can provide a system-level insight into genotype-phenotype relationships in fermentative communities.

The field remains largely unexplored despite some advancements in our understanding of the molecular dynamics from a transcriptional perspective, particularly in fermented food products. Although global omics parallel profiling of transcripts and proteins has several clear advantages, there are biological and practical differences in how they are used. Transcriptomics can now measure tens of thousands of specific mRNA species at once and is becoming inexpensive. However, lack of consistency between sequencing technologies and inconsistent read depth (the percentage of total transcripts sequenced) may affect repeatability.

9.3 Metabolomics

Metabolomics is the comprehensive analysis of small-molecule metabolites within a biological sample, which can aid comprehension of the general metabolic alterations brought about by fermentation (Zhan et al., 2023). Metabolomics uses both global untargeted metabolomics and targeted metabolomics. Untargeted metabolomics is a thorough analytical method designed to identify and examine every metabolite found in a particular biological system impartially. It is mostly used for the discovery of the unknown, enabling scientists to find new, previously unidentified molecules and metabolites. Such discoveries can lead to a new understanding of the fermentation process and the uniqueness of fermented foods. For efficient data mining and interpretation, untargeted metabolomics necessitates the integration of chemometric techniques due to the complexity and volume of data it creates. Usually qualitative, semiquantitative and data-driven, this method is frequently employed to produce rather than test preconceived hypotheses (Adebo et al., 2021; Roberts et al., 2012).

Conversely, targeted metabolomics concentrates on a particular group of recognised metabolites, frequently associated with specific biochemical pathways or functional characteristics in foods; for example, amino acids for nutritional profiling or phenolics for health benefits. This method depends on absolute measurement using established chemical standards and usually entails hypothesis testing. Targeted metabolomics is more structured, requiring the selection of metabolites of interest before examination and employing particular, frequently more sensitive analytical techniques for accurate quantification, even if it can still be somewhat data-driven (Lacalle-Bergeron et al., 2021). Analytical platforms for metabolomics for both targeted and untargeted approaches include nuclear magnetic resonance (NMR) spectroscopy, GC-MS, liquid chromatography–mass spectrometry (LC–MS) and capillary electrophoresis–mass spectrometry (CE-MS).

Metabolomics is rapidly emerging as a powerful tool for explicating the biochemical complexity of traditional African fermented foods. In the African context, metabolomics is instrumental in identifying key bioactive compounds, including short-chain fatty acids, polyphenol derivatives and microbial secondary metabolites, all of which contribute to the sensory, nutritional and functional attributes of fermented products. Adebo et al. (2019) employed metabolomics to unravel the bioactive compounds in a fermented sorghum-based African porridge (*ting*) using *Lactobacillus fermentum*. Their study applied GC–MS to decipher the intricate biochemical changes that occur during fermentation, providing information about potential health benefits and nutritional quality (Adebo et al., 2019). In fermented maize gruel (*ogi*), certain groups of metabolites are reportedly prominent, such as alcohols, aldehydes, ergot alkaloids, fatty acids, furanones, indole and pyridines (Odukoya et al., 2023). In research focusing on *mahewu*, a fermented maize-based beverage, researchers investigated how different microbial starters affect its metabolite profile. The findings revealed that the choice of starter culture significantly impacts the nutritional profiles and sensory qualities of *mahewu*, with variations observed in compounds such as phenols, fatty acids and vitamins (Akanni et al., 2025; Daji et al., 2024). Metabolomics has the potential to facilitate the assessment of fermentation consistency and detection of spoilage-related compounds in African fermented foods and, thus, the authentication of region-specific fermentation practices. Despite its potential, the application of metabolomics in African fermented foods remains limited by infrastructural and analytical constraints. The possibility of strengthening regional metabolomics capacity through interdisciplinary collaboration can drive innovation and improve the food safety, quality and commercialisation of these products in Africa, which would enable a comprehensive, systems-level understanding of African fermented food ecosystems.

9.4 Proteomics

Proteomics is the large-scale study of proteins, including their structures, functions and interactions within a biological system at a given juncture (Pandey & Mann, 2000). In the context of fermented products, it refers to the comprehensive study and characterisation of the entire set of proteins (the proteome) expressed by microbial communities and the food matrix during fermentation (Chen et al., 2017; Yang et al., 2020). Techniques commonly employed are two-dimensional (2D) acrylamide gel electrophoresis and mass spectrometry, such as matrix-assisted laser desorption/ionisation time-of-flight mass spectrometry (MALDI-TOF-MS), liquid chromatography coupled to tandem mass spectrometry (LC–MS/MS) or quadrupole time-of-flight mass spectrometry (Q-TOF-MS).

Recent advances in high-throughput proteomics, especially tandem mass spectrometry (MS/MS), have enabled detailed profiling of microbial and plant proteins in fermented foods, revealing enzymatic changes, protein breakdown and bioactive peptide formation. Studies on cereal-based fermentation have shown extensive proteolysis-producing peptides with antihypertensive and antioxidant potential (Abrahams, 2023). A primary use of proteomics in the study of African fermented foods is to identify surface polypeptides of the microbial communities present. Soro-Yao and coworkers (2014) used MALDI-TOF mass spectrometry ribotyping to identify LAB from fermented cereal foods in Abidjan (Côte d'Ivoire) (Soro-Yao et al., 2014). MALDI-TOF profiling has identified LAB in *kunu-zaki* (cereal-based fermented beverage), *kunu aya* (tiger nut beverage) and *Bacillus* species from *dawadawa* (condiment), all fermented products from Nigeria (Akanni, Naudé, et al., 2018; Ogunremi et al., 2022, 2024).

9.5 Multi-Omics Integration to Understand African Fermented Foods

The combination of advanced high-throughput omics technologies has transformed the field of biological research. Technologies such as (meta) genomics, (meta) transcriptomics, (meta) proteomics, metabolomics and metagenomics allow for in-depth investigation of biological processes on several molecular levels. Multi-omics integration, the sequencing of two or more omics layers, has proven to be an effective strategy for uncovering microbial interactions, metabolic pathways and dynamic functional responses (Misra et al., 2018; Wen et al., 2024).

African fermented foods, including *ogi, nunu, kenkey, ugba* and *injera*, have long-standing cultural and nutritional importance. Fermentation in the production of these foods is predominantly spontaneous or employs traditional back-slopping and is dependent on complex and poorly defined microbial populations. Despite some available microbiological characterisations, much of the published literature centres on 16S rRNA profiling or culture-dependent methods (Adewumi et al., 2012; Cocolin et al., 2013; Diaz et al., 2019). These approaches have identified core microbial taxa, including LAB strains (e.g., *Lactobacillus*, *Leuconostoc*), yeasts (*Saccharomyces, Candida*) and *Bacillus* spp., but not the functional roles of these microbes in fermentation. This functional deficiency can be overcome by applying a multi-omics approach. Metagenomics offers valuable insights into the taxonomic origin and genetic potential of microbial communities; however, coupling this with metatranscriptomics or metaproteomics identifies active metabolic pathways and regulatory mechanisms (Jansson & Baker, 2016). Metabolomics contributes to the overall understanding by characterising the biochemical end products of fermentation, including organic acids, alcohols, peptides and volatile molecules that impact flavour and nutritional value (Pinu et al., 2019). Together, all these elements enable a comprehensive understanding of the fermentation processes.

Despite the extensive use of multi-omics in Asian fermented foods, such as *kimchi* and cheese (Chun et al., 2017; Jung et al., 2013; Lessard et al., 2014), its use in African fermented foods has been limited. The majority of research on African foods is confined to taxonomic profiling, and not many have incorporated functional or molecular data. For example, some studies have described the bacterial and fungal populations in fermented foods from Africa through 16S rRNA sequencing (Diaz et al., 2019; Hlangwani et al., 2023; Oguntoyinbo et al., 2011), but there is a paucity of studies investigating real-time gene expression or metabolite flux during fermentation. This deficit represents a substantial untapped potential to understand how microbial activities unfold during fermentation and impact food quality and safety.

Initiatives such as the African Fermented Foods Biobank systematically document microbial strains and their characteristics. Several recent publications employed a combined approach of metagenomics and metabolomics to study products such as *amasi* (fermented milk) and *amahewu* (fermented beverage), thereby demonstrating initial correlations between microbial taxa and metabolite production (Ajibade et al., 2024; Kgoale, 2023). However, significant challenges remain, such as poor access to sequencing infrastructure, insufficient computational power and underrepresentation of African food microbiomes in international databases. Thus, the integration of multi-omics has tremendous potential for scientific breakthroughs as it would be capable of identifying crucial microbial taxa, elucidating metabolite biosynthetic processes and facilitating predictive modelling of fermentation.

9.6 Conclusions

African food products and fermentation practices are increasingly being revisited through the lens of advanced omics technologies. Tools such as transcriptomics, metabolomics and proteomics, alongside integrative multi-omics approaches, are providing insights into the microbial communities, biochemical processes and functional properties that define these foods. While the application of omics in African fermented foods is still emerging, initial studies demonstrate its vast potential. These technologies provide an understanding of microbial dynamics, gene expression, metabolite pathways and protein interactions, which are instrumental in improving food safety, nutritional quality and product standardisation. Despite infrastructural challenges and limited representation in global microbiome databases, growing interest in the integration of multi-omics platforms represents a frontier that can transform African fermented food systems. Therefore, future research should focus on comprehensive, locally driven and systems-level studies that use omics technologies in an inclusive, accessible and sustainable manner.

References

Abrahams, A. M. (2023). Proteomics and transcriptomics and their application in fermented foods. In *Indigenous fermented foods for the tropics* (pp. 377–391). Elsevier.

Adebo, O. A., Kayitesi, E., Tugizimana, F., & Njobeh, P. B. (2019). Differential metabolic signatures in naturally and lactic acid bacteria (LAB) fermented ting (a Southern African food) with different tannin content, as revealed by gas chromatography mass spectrometry (GC–MS)-based metabolomics. *Food Research International*, *121*, 326–335.

Adebo, O. A., Oyeyinka, S. A., Adebiyi, J. A., Feng, X., Wilkin, J. D., Kewuyemi, Y. O., Abrahams, A. M., & Tugizimana, F. (2021). Application of gas chromatography–mass spectrometry (GC-MS)-based metabolomics for the study of fermented cereal and legume foods: A review. *International Journal of Food Science and Technology*, *56*(4), 1514–1534.

Adewumi, G. A., Oguntoyinbo, F. A., Keisam, S., Romi, W., & Jeyaram, K. (2012). Combination of culture-independent and culture-dependent molecular methods for the determination of bacterial community of iru, a fermented *Parkia biglobosa* seeds. *Frontiers in Microbiology*, *3*, 436. doi:10.3389/fmicb.2012.00436

Ajibade, B. O., Ajayeoba, T. A., Sabiu, S., Moiseenko, K. V., Mbona, S. V., Cason, E. D., Fedorova, T. V., & Ijabadeniyi, O. A. (2024). Unveiling the microbial symphony of amasi: A targeted metagenomic 16S rRNA, ITS, and metabolites insights using bovine and caprine milk. *Fermentation*, *11*(1). doi:10.3390/fermentation11010006

Akanni, G. B., Daji, G. A., Green, E., & Adebo, O. A. (2025). Comparative metabolomics investigation of white maize mahewu: Understanding biochemical transformations induced by traditional cereal-based starters. *International Journal of Food Science and Technology*, *60*(1), vvaf044.

Akanni, G. B., De Kock, H. L., Naudé, Y., & Buys, E. M. (2018). Volatile compounds produced by *Bacillus* species alkaline fermentation of bambara groundnut (*Vigna subterranean* (L.) Verdc) into a dawadawa-type

African food condiment using headspace solid-phase microextraction and GC× GC–TOFMS. *International Journal of Food Properties*, *21*(1), 930–942.

Akanni, G. B., Naudé, Y., de Kock, H. L., & Buys, E. M. (2018). Diversity and functionality of bacillus species associated with alkaline fermentation of bambara groundnut (*Vigna subterranean* L. Verdc) into dawadawa-type African condiment. *European Food Research and Technology*, *244*, 1147–1158.

Azokpota, P., Hounhouigan, J. D., Annan, N. T., Nago, M. C., & Jakobsen, M. (2008). Diversity of volatile compounds of afitin, iru and sonru, three fermented food condiments from Benin. *World Journal of Microbiology and Biotechnology*, *24*, 879–885.

Chen, G., Chen, C., & Lei, Z. (2017). Meta-omics insights in the microbial community profiling and functional characterization of fermented foods. *Trends in Food Science & Technology*, *65*, 23–31.

Chun, B. H., Kim, K. H., Jeon, H. H., Lee, S. H., & Jeon, C. O. (2017). Pan-genomic and transcriptomic analyses of *Leuconostoc mesenteroides* provide insights into its genomic and metabolic features and roles in kimchi fermentation. *Scientific Reports*, *7*(1), 11504. doi:10.1038/s41598-017-12016-z

Cifuentes, A. (2009). *Food analysis and foodomics.*

Cocolin, L., Alessandria, V., Dolci, P., Gorra, R., & Rantsiou, K. (2013). Culture independent methods to assess the diversity and dynamics of microbiota during food fermentation. *International Journal of Food Microbiology*, *167*(1), 29–43. doi:10.1016/j.ijfoodmicro.2013.05.008

Conesa, A., Madrigal, P., Tarazona, S., Gomez-Cabrero, D., Cervera, A., McPherson, A., Szcześniak, M. W., Gaffney, D. J., Elo, L. L., Zhang, X., & Mortazavi, A. (2016). A survey of best practices for RNA-seq data analysis. *Genome Biology*, *17*, 13. doi:10.1186/s13059-016-0881-8

Daji, G. A., Green, E., Akanni, G. B., & Adebo, O. A. (2024). GC–MS-based metabolomics elucidates the impact of cereal-based microbial starters on the metabolite profile of mahewu prepared from yellow maize. *International Journal of Food Science and Technology*, *59*(6), 4268–4280.

Dhellot, J., Mokemiabeka, S., Moyen, R., Kobawila, S., & Louembe, D. (2014). Volatile compounds produced in two traditional fermented foods of the Congo: Nsamba (palm wine) and bikedi (retted cassava dough). *African Journal of Biotechnology*, *13*(42).

Diaz, M., Kellingray, L., Akinyemi, N., Adefiranye, O. O., Olaonipekun, A. B., Bayili, G. R., …, & Narbad, A. (2019). Comparison of the microbial composition of African fermented foods using amplicon sequencing. *Scientific Reports*, *9*(1), 13863. doi:10.1038/s41598-019-50190-4

Gadaga, T. H., Viljoen, B. C., & Narvhus, J. A. (2007). Volatile organic compounds in naturally fermented milk and milk fermented using yeasts, lactic acid bacteria and their combinations as starter cultures. *Food Technology and Biotechnology*, *45*(2), 195–200.

Hlangwani, E., Abrahams, A., Masenya, K., & Adebo, O. A. (2023). Analysis of the bacterial and fungal populations in South African sorghum beer (umqombothi) using full-length 16S rRNA amplicon sequencing. *World Journal of Microbiology and Biotechnology*, *39*(12), 350. doi:10.1007/s11274-023-03764-4

Jansson, J. K., & Baker, E. S. (2016). A multi-omic future for microbiome studies. *Nature Microbiology*, *1*, 16049. doi:10.1038/nmicrobiol.2016.49

Jung, J. Y., Lee, S. H., Jin, H. M., Hahn, Y., Madsen, E. L., & Jeon, C. O. (2013). Metatranscriptomic analysis of lactic acid bacterial gene expression during kimchi fermentation. *International Journal of Food Microbiology*, *163*(2-3), 171–179. doi:10.1016/j.ijfoodmicro.2013.02.022

Kgoale, S. D. P. (2023). *Bacterial microbiota and metabolomic profiling of amahewu production*. University of Johannesburg (South Africa.

Lacalle-Bergeron, L., Izquierdo-Sandoval, D., Sancho, J. V., Lopez, F. J., Hernandez, F., & Portoles, T. (2021). Chromatography hyphenated to high resolution mass spectrometry in untargeted metabolomics for investigation of food (bio) markers. *TrAC Trends in Analytical Chemistry*, *135*, 116161.

Lessard, M. H., Viel, C., Boyle, B., St-Gelais, D., & Labrie, S. (2014). Metatranscriptome analysis of fungal strains *Penicillium camemberti* and *Geotrichum candidum* reveal cheese matrix breakdown and potential development of sensory properties of ripened Camembert-type cheese. *BMC Genomics*, *15*, 235. doi:10.1186/1471-2164-15-235

Li, S., Tian, Y., Jiang, P., Lin, Y., Liu, X., & Yang, H. (2021). Recent advances in the application of metabolomics for food safety control and food quality analyses. *Critical Reviews in Food Science and Nutrition*, *61*(9), 1448 1469.

Misra, B. B., Langefeld, C. D., Olivier, M., & Cox, L. A. (2018). Integrated omics: Tools, advances, and future approaches. *Journal of Molecular Endocrinology*. doi:10.1530/JME-18-0055

Oboh, G., & Akindahunsi, A. (2003). Biochemical changes in cassava products (flour & gari) subjected to *Saccharomyces cerevisae* solid media fermentation. *Food Chemistry*, *82*(4), 599–602.

Oboh, G., Akindahunsi, A., & Oshodi, A. (2002). Nutrient and anti-nutrient contents of *Aspergillus niger*-fermented cassava products (flour and gari). *Journal of Food Composition and Analysis*, *15*(5), 617–622.

Odukoya, J. O., De Saeger, S., De Boevre, M., Adegoke, G. O., Devlieghere, F., Croubels, S., Antonissen, G., Adebo, O. A., Gbashi, S., Odukoya, J. O., & Njobeh, P. B. (2023). Mycotoxin reduction and metabolite profiles of ogi produced using traditional fermentation methods. *Food Hydrocolloids for Health*, *4*, 100160.

Ogunremi, O. R., Freimuller Leischtfeld, S., & Miescher Schwenninger, S. (2022). MALDI-TOF MS profiling and exopolysaccharide production properties of lactic acid bacteria from Kunu-zaki - A cereal-based Nigerian fermented beverage. *International Journal of Food Microbiology*, *366*, 109563. doi:10.1016/j.ijfoodmicro.2022.109563

Ogunremi, O. R., Ganz, G., Freimüller Leischtfeld, S., & Miescher Schwenninger, S. (2024). MALDI-TOF MS profiling and antifungal activity of lactic acid bacteria from kunu aya, a tiger nut traditional beverage of Nigeria. *Food Bioscience*, *61*. doi:10.1016/j.fbio.2024.104581

Oguntoyinbo, F. A., Tourlomousis, P., Gasson, M. J., & Narbad, A. (2011). Analysis of bacterial communities of traditional fermented West African cereal foods using culture independent methods. *International Journal of Food Microbiology*, *145*(1), 205–210. doi:10.1016/j.ijfoodmicro.2010.12.025

Olapade, A., Babalola, Y., & Aworh, O. (2014). Quality attributes of fufu (fermented cassava) flour supplemented with bambara flour. *International Food Research Journal*, *21*(5).

Ouoba, L., Diawara, B., Annan, N., Poll, L., & Jakobsen, M. (2005). Volatile compounds of Soumbala, a fermented African locust bean (*Parkia biglobosa*) food condiment. *Journal of Applied Microbiology*, *99*(6), 1413–1421.

Pandey, A., & Mann, M. (2000). Proteomics to study genes and genomes. *Nature*, *405*(6788), 837–846.

Pinu, F. R., Beale, D. J., Paten, A. M., Kouremenos, K., Swarup, S., Schirra, H. J., & Wishart, D. (2019). Systems biology and multi-omics integration: Viewpoints from the metabolomics research community. *Metabolites*, *9*(4). doi:10.3390/metabo9040076

Roberts, L. D., Souza, A. L., Gerszten, R. E., & Clish, C. B. (2012). Targeted metabolomics. *Current Protocols in Molecular Biology*, Chapter 30: Unit 30.2. 1–24. doi: 10.1002/0471142727

Satam, H., Joshi, K., Mangrolia, U., Waghoo, S., Zaidi, G., Rawool, S., Thakare, R. P., Banday, S., Mishra, A. K., Das, G., & Malonia, S. K. (2023). Next-generation sequencing technology: Current trends and advancements. *Biology (Basel)*, *12*(7). doi:10.3390/biology12070997

Soro-Yao, A. A., Schumann, P., Thonart, P., Djè, K. M., & Pukall, R. (2014). The use of MALDI-TOF mass spectrometry, ribotyping and phenotypic tests to identify lactic acid bacteria from fermented cereal foods in Abidjan (Côte d'Ivoire). *The Open Microbiology Journal*, *8*, 78.

Sripathi, V. R., Anche, V. C., Gossett, Z. B., & Walker, L. T. (2021). Recent applications of RNA sequencing in food and agriculture. In *Applications of RNA-seq in biology and medicine*: IntechOpen.

Tamang, J. P., Cotter, P. D., Endo, A., Han, N. S., Kort, R., Liu, S. Q., Mayo, B., Westerik, N., & Hutkins, R. (2020). Fermented foods in a global age: East Meets West. *Comprehensive Reviews in Food Science and Food Safety*, *19*(1), 184–217. doi:10.1111/1541-4337.12520

Turpin, W., Humblot, C., Thomas, M., & Guyot, J.-P. (2010). Lactobacilli as multifaceted probiotics with poorly disclosed molecular mechanisms. *International Journal of Food Microbiology*, *143*(3), 87–102.

Wang, Z., Gerstein, M., & Snyder, M. (2009). RNA-seq: A revolutionary tool for transcriptomics. *Nature Reviews Genetics*, *10*(1), 57–63.

Wen, L., Yang, L., Chen, C., Li, J., Fu, J., Liu, G., Kan, Q., Ho, C.T., Huang, Q., Lan, Y., & Cao, Y. (2024). Applications of multi-omics techniques to unravel the fermentation process and the flavor formation mechanism in fermented foods. *Critical Reviews in Food Science and Nutrition, 64*(23), 8367–8383. doi:10.1080/10408398.2023.2199425

Yang, H., He, Y., Liao, J., Li, X., Zhang, J., Liebl, W., & Chen, F. (2022). RNA-Seq transcriptomic analysis reveals Gene expression profiles of acetic acid bacteria under high-acidity submerged industrial fermentation process. *Frontiers in Microbiology, 13*, 956729. doi:10.3389/fmicb.2022.956729

Yang, L., Fan, W., & Xu, Y. (2020). Metaproteomics insights into traditional fermented foods and beverages. *Comprehensive Reviews in Food Science and Food Safety, 19*(5), 2506–2529.

Zhan, Q., Thakur, K., Feng, J.-Y., Zhu, Y.-Y., Zhang, J.-G., & Wei, Z.-J. (2023). LC–MS based metabolomics analysis of okara fermented by *Bacillus subtilis* DC-15: Insights into nutritional and functional profile. *Food Chemistry, 413*, 135656.

10

Conclusion and Future Perspectives

As frequently highlighted in this book, African fermented foods are spread across the continent and continue to contribute to the cultural heritage and diet of the African populace. The unique, intricate microbial diversity and systems that underpin the fermentation of cereals (Chapter 2), legumes, pulses and oil seeds (Chapter 3), dairy (Chapter 4), roots and tubers (Chapter 5), fish and meat (Chapter 6), fruits and vegetables (Chapter 7) have been explored. Existing analytical techniques that assist in examining their microbiota as well as evolving ones that could further deepen our understanding of these products have also been examined (Chapters 8 and 9). Modern omics tools like metagenomics, proteomics and metabolomics are revolutionising our understanding of microbial diversity, succession and function. Recent incorporation of multi-omics approach can uncover gut-health benefits and probiotic potential, emphasising the need for interdisciplinary efforts that merge science with traditional knowledge of African fermented foods (Ferrocino et al., 2023; Zhang et al., 2025).

African fermented foods stand at a transformative crossroads – convergence of traditional knowledge systems with cutting-edge food biotechnology to address contemporary 21st-century challenges of food security, sustainable development and nutrition. With a projected global fermented foods revenue sales of US$312.04 billion by 2033 and Africa's share of US$54.61 billion (Fermented Foods Market Report, 2025), this market is expected to continually increase. This is linked to the popularity of fermented foods due to their specific health benefits (Adebo, 2025), consumer priorities on wellness as well as their cultural significance, especially in Africa. Unprecedented developments have also occurred over the years with these African fermented foods transitioning from subsistence-based production confined to local markets to improved value chains, supporting the continent's food

DOI: 10.1201/9781003658320-10

security objectives and meeting international standards. The latter is in fact demonstrated in the availability of some notable African fermented food products in international stores accessible to migrants who might not have primarily had access to them.

Economically, production of African fermented food products provides crucial employment opportunities throughout the value chain, particularly for women who are usually much more involved and traditionally manage fermentation processes across the continent. Much more could, however, be done in addressing infrastructural challenges that could be associated with processing of fermented foods (Sobowale et al., 2023) and development of market-ready products. Furthermore, creating enabling conditions, appropriate investment and technical support and access to processing technologies through different nodes, hubs and possibly governmental agencies will further assist in ensuring that these traditional foods achieve significant market penetration.

Safety of fermented foods has been a concern with studies highlighting contamination with pathogenic microorganisms, biogenic amines and mycotoxins (Anyogu et al., 2021; El-Imam, 2023; Hawaz et al., 2025; Omidiran & Jenfa, 2023; Owusu-Kwarteng et al., 2020; Oyedeji et al., 2023). These are often linked to contaminated raw materials/substrates, poor hygiene and lack of standardised process, packaging and storage. To thus effectively address these issues, it is vital to develop and implement hazard analysis and critical control points (HACCP) systems and good manufacturing practices (GMP) tailored to traditional food settings, regular monitoring for common and emerging food contaminants as well as the training of producers and processors. Improving the safety of these products will protect consumers but also enhance the export potential of these African fermented food products.

There is also the need to leverage on synergistic technologies that would benefit African fermented foods (Figure 10.1). These include but not limited to innovative starter culture development (single-strain, multi-strain or multi-strain mixed indigenous cultures) that can replicate the complexity of spontaneous fermentation for African fermented food products (Adebo et al., 2018) as well as fermentation monitoring technologies that could track fermentation processes in real time (Adeleke et al., 2023; Adeleke et al., 2025) and be applicable

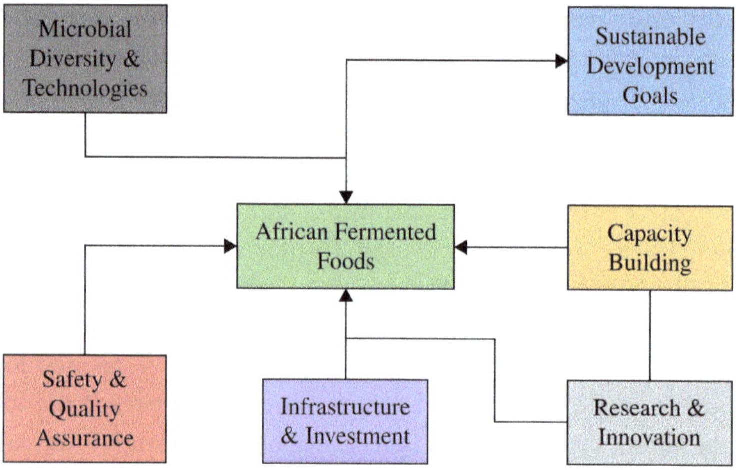

Figure 10.1 Integrative framework for advancing African fermented foods through synergistic approaches.

in resource-limited environments, providing reliable process control data. It could be opined that the future of African fermented foods lies in the intelligent integration of traditional knowledge with appropriate modern technologies.

The projection for African fermented foods in the nearest future is promising. Much more is, however, needed to be done including deliberate investment in research and innovation to improve product quality and deepen our understanding of the microbiology and other vital aspects of African fermented foods. Vital here is also necessary capacity building through education, infrastructure development and training to support the value chain of African fermented foods. Associated support through the aforementioned and more by the relevant stakeholders is necessary to contribute to the resilience of African fermented foods.

References

Adebo, O. A. (2025). Discovering health-promoting effects of fermented food: From diabetes prevention to anti-cancer possibilities. *Sight and Life Special Report: The Science*, 32–37.

Adebo, O. A., Njobeh, P. B., Adeboye, A. S., Adebiyi, J. A., Sobowale, S. S., Ogundele, O. M., & Kayitesi, E. (2018). Advances in fermentation technology for novel food products. In Panda, S., & Shetty, P. (Eds.),

Innovations in technologies for fermented food and beverage industries (pp. 71–87). Springer.

Adeleke, I., Nwulu, N., & Adebo, O. A. (2023). Internet of things (IoT) in the food fermentation process: A bibliometric review. *Journal of Food Process Engineering, 46*, e14321.

Adeleke, I., Nwulu, N., & Adebo, O. A. (2025). Leveraging IoT and machine learning for smart fermentation of *amasi*: A predictive framework for acidity control. *Journal of Agriculture and Food Research, 24*, 102409.

Anyogu, A., Olukorede, A., Anumudu, C., Onyeaka, H., Areo, E., Obadina, A., Odimba, J. N., & Nwaiwu, O. (2021). Microorganisms and food safety risks associated with indigenous fermented foods from Africa. *Food Control, 129*, 108227. https://doi.org/10.1016/j.foodcont.2021.108227

El-Imam, A. A. (2023). Occurrence of mycotoxins in fermented tropical foods. In Adebo, O. A., Chinma, C. E., Obadina, O., Panda, S., Soares, A., Gan, R. Y. (Eds.), *Indigenous fermented foods in the tropics* (pp. 505–517). Elsevier. https://doi.org/10.1016/B978-0-323-98341-9.00019-0

Fermented Foods Market Report 2025 (Global Edition) (2025). Available from: https://www.cognitivemarketresearch.com/fermented-foods-market-report (Accessed 26/06/2025).

Ferrocino, I., Rantsiou, K., McClure, R., Kostic, T., de Souza, R. S. C., Lange, L., FitzGerald, J., Kriaa, A., Cotter, P., & Maguin, E. (2023). The need for an integrated multi-OMICs approach in microbiome science in the food system. *Comprehensive Reviews in Food Science and Food Safety, 22*, 1082–1103. https://doi.org/10.1111/1541-4337.13103

Hawaz, H., Bottari, B., Scazzina, F., & Carini, E. (2025). Eastern African traditional fermented foods and beverages: Advancements, challenges, and perspectives on food technology, nutrition, and safety. *Comprehensive Reviews in Food Science and Food Safety, 24*, e70137. https://doi.org/10.1111/1541-4337.70137

Omidiran, A. T., & Jenfa, M. D. (2023). Occurrence of biogenic amines in fermented foods. In Adebo, O. A., Chinma, C. E., Obadina, O., Panda, S., Soares, A., & Gan, R. Y. (Eds.). *Indigenous fermented foods in the tropics* (pp. 539–548). Elsevier. https://doi.org/10.1016/B978-0-323-98341-9.00027-X

Owusu-Kwarteng, J., Akabanda, F., Agyei, D., & Jespersen, L. (2020). Microbial safety of milk production and fermented dairy products in Africa. *Microorganisms, 8*, 752. https://doi.org/10.3390/microorganisms 8050752

Oyedeji, A. B., Green, E., Jeff-Agboola, Y. A., Olanbiwoninu, A. A., Areo, E., Martins, I. E., El-Imam, A. M. A., & Adebo, O. A. (2023). Presence of pathogenic microorganisms in fermented foods. In Adebo, O. A., Chinma, C. E., Obadina, O., Panda, S., Soares, A., & Gan, R. Y.(Eds.), *Indigenous fermented foods in the tropics* (pp. 519–537). Elsevier. https://doi.org/10.1016/B978-0-323-98341-9.00037-2

Sobowale, S. S., Olatidoye, O. P., Omosebi, M. O., & Agbawodike, J. I. (2023). Equipment and machinery for improving the fermentation process of indigenous foods. In Adebo, O. A., Chinma, C. E., Obadina, O., Panda, S., Soares, A., Gan, R. Y. (Eds.), *Indigenous fermented foods in the tropics* (pp. 433–465). Elsevier. https://doi.org/10.1016/B978-0-323-98341-9.00034-7

Zhang, Y., Zhu, X., Wang, N., Liu, X., Wang, L., & Ning, K. (2025). Synergy of traditional practices and modern technology: Advancing the understanding and applications of microbial resources and processes in fermented foods. *Trends in Food Science & Technology*, *157*, 104891. https://doi.org/10.1016/j.tifs.2025.104891

Index

Note: Page numbers in *italics* and **bold** refer to figures and tables.

L

M

N

O

P

Q

R

For Product Safety Concerns and Information please contact our EU representative GPSR@taylorandfrancis.com
Taylor & Francis Verlag GmbH, Kaufingerstraße 24, 80331 München, Germany

www.ingramcontent.com/pod-product-compliance
Ingram Content Group UK Ltd.
Pitfield, Milton Keynes, MK11 3LW, UK
UKHW022211190926
475702UK00003B/299
9781041111085